The Nature of Energy

Patrick Boardman

Published by Patrick Boardman, 2020.

THE NATURE OF ENERGY

First edition. March 5, 2020.

ISBN: 979-8215375068

Written by Patrick Boardman.

Preface

This book explains the facts on how Albert Einstein came to write the famous formula that led to the Atomic Age and verifies Max Planck's hypothesis on the nature of matter. Planck stated that a conscious force is the matrix of all matter. Like most people, I spent decades giving too much credit for existence to matter, for in reality all material things are vibrating fields of highly concentrated energy. There has been much conjecture put forward about the mechanisms of our very existence, but writers and speakers offer ideas based on assumptions rather than presenting evidence that leads to a proper chain of logic.

My conclusions are the result of fifty-seven years of research. In this book I will reveal new information on the Big Bang of the universe and explain the correct model of material and non-material existence. Light is a wave that collapses into particle form once observed, and Quantum Theory also sees matter as a wave phenomenon that experiences solidity through consciousness.

Light is a transverse electromagnetic wave that can been seen by the receptors in our eyes. An expanding sphere of light behaves as if each point on the wave front were a new source of radiation of the same frequency and phase. The speed of light in a vacuum is fixed at 299,792,458 meters per second by the current definition of the meter. Light's velocity is expressed in physics by the letter 'c' which stands for the Latin word for swiftness, '*celeritas*'. The speed of light in a vacuum is a universal constant in all reference frames. The universe had a starting point, as did light. We perceive the material dimension, but there is more to existence than the universe, which is finite and has borders. Energy has no borders, nor does it have an end point, since energy cannot be created or destroyed. Power is the harnessing of potential energy for transmission or storage, so we must use the two terms separately when discussing the principles of energy. Energy can be concentrated but it cannot be dissipated into oblivion.

Light is also known as visible light to differentiate it from 'ultraviolet light' and 'infrared light'. The frequency of a light wave is relative to its color. Monochromatic light, such as Laser light, is described by only one frequency. Six colors are associated with a band of monochromatic light. In order of increasing frequency, they are red, orange, yellow, green, blue, and violet. Polychromatic light is described by many different frequencies. This is the white light we see every day that allows us to see a wide range of colors. The intensity versus the frequency produces the colors we know as a spectrum.

There are three fundamental quantities of nature: Energy, Mass, and Momentum. We see Einstein's formula $E= mc^2$, revealing that matter is concentrated Energy, but to account for the effects of Momentum, the formula is correctly written as the Relativistic Dispersion Relation: $E^2= m^2 c^4+ p^2 c^2$

The state of being alive as a conscious individual has always seemed miraculous to me, but many thoughtless people consider life an accident without any meaning. Philosophical minds agree that life is a precious gift that should not be taken for granted.

The Unified Field Theory addresses certain philosophical concepts gleaned from the various belief systems of the world. Many people will dismiss any ideas that conflict with their traditional beliefs so they will react with hostility towards new information that goes against religious dogma. Humans cannot bring themselves to admit they are wrong about anything.

The oversized human ego is at the heart of many misunderstandings. Thousands of different religions have developed over the ages. Conflicts between doctrines led to violence and strife. Traditional beliefs and practices have corrupted the major religions, turning many practitioners into hostile warring parties determined to prevail over the others. Sadly, people insist on being the judge of others and they are willing to go to war to prove they are right. Even atheists will become angry when they are challenged or debated, as they strive to defend their belief in nothing.

The main purpose of this book is to present scientific proof of God. Some people will remain unconvinced no matter how much evidence is given, but those with open minds will now have the information that is necessary to save themselves from punishment in this life and the prospect of existing in pain, fear, and loneliness for all eternity. Denying that there is Divine Retribution will not stop the punishment from happening – whether it is retribution by karma within our lifetimes or permanent punishment after Judgment Day, or both. All people should strive to attain Heaven, where the souls who are pure and righteous will experience only reward and pleasure forever in the Afterlife. Those who do not meet the requirements for Paradise will exist in the opposite state, where there will be never ending misery and humiliation. Current events are proof that mankind could be wiped out at any time at the whim of crazed warmongers.

The Einstein Project

New concepts in theorical Physics often extend beyond university blackboards to the point where they have a profound effect on society one way or the other, such as when a vital clue was published in 1905 by a well-known pacifist Professor Albert Einstein, which eventually led to the development of the atomic bomb. Leo Szilard was a Hungarian-German physicist who discovered the nuclear chain reaction in 1933 and patented the first nuclear fission reactor. He wrote up a letter for Einstein's signature to send to Franklin Roosevelt's attention, warning the US president of the possibility of an atomic bomb. The letter was taken seriously, and it led to the start of the Manhattan Project, which shortened the war against Japan.

The history of mankind has been one of continuous conflict between religions, ethnic groups, and economic classes. There are certain truths that all people should follow for reasons I will outline in each chapter. In the decade after the Second World War mankind was awakened to the reality of a possible nuclear war, and so there was increased interest in science among the young people who were confronted by the threat of sudden annihilation. Physics had proven to be an immensely powerful and serious area of study. I was taught to read at home, so I had a great advantage when I started attending school. My first teacher was an elderly nun at one of the schools operated by Notre Dame University. Sister Eustice was considered a genius by her peers, and she held many doctorates. She took me aside after class one day and asked me to visit the library and study all the writings of Albert Einstein, who had died without finishing his work. The scholars needed someone young because the project would take a lifetime to complete. The wise old nun then told me the secret of why I was chosen to do this work. Readers familiar with the Dead Sea Scrolls and the term "The Book that will never decay" will be able to deduce that secret. The things Sister Eustice told me caused me to quit the Catholic church as soon as I could. The Vatican ruled Europe for centuries using murder, war, and torture to promote its goal of total world control by the pope of Rome. Those who commit heinous crimes of violence are not carrying out an admirable spiritual campaign. Catholicism is a cult with pagan roots and traditions who insist that the only people going to Heaven are its members. The papal ideal is to establish one world government, one police force, one army, and only one common religion...the universal religion that will dominate all nations. The method of obtaining power is infiltration of the Jesuit Army of Loyola into all facets of society.

The world is experiencing turmoil again, although the nature of the threat is a matter of opinion and debate among divided groups. The scientific method asks us to observe everything neutrally as outside analysts who are open minded. In my years of research, I found valuable connections and clues in other areas of study, such as History, Biology, Theology, and Philosophy. My task is to solve this problem: is there a meaning to life, or is life accidental and totally meaningless? There are those who maintain that nothing erupted into everything spontaneously and all things are random flukes of nature. Others believe in one or more divine beings, while the new age movement claims that we are all part of one big consciousness, destined to return to oneness at some point. Atheists and religious people are not likely to change, for they do not have open minds. Many people agree that governments are being oppressive at a time when we should have developed an ideal way of life full of peace and happiness. The questionable two year lockdown was a setback for some people, a disaster for others. History shows that similar events in modern medicine took place in 1918, when a citizen could be sent to jail for failing to wear a cloth mask in public. The gullible public has always been conditioned to take orders from the powers that be, so they continue to cooperate in the many fear campaigns launched by their elite rulers. Officials and corporations share in the plunder gained in fraudulent activities against the people. Clearly, they do not believe they will ever be punished for their greed. Even proof of a Supreme Being will not deter them, they will simply laugh it off. I doubt that there will ever be any fear of God in them, for the criminals are obsessed with the love of money. The first step in proving an equation is to perform a verification, but it became apparent from the outset that I could not verify Einstein's equation, and neither could anybody else in the scientific community. I corresponded by mail with many experts in Physics to ask advice, but no one knew how to do a verification. Einstein never told anyone how he got the formula, and his writings had no background work that would lead to his conclusion that mass equals Energy divided by the speed of light squared. The formula that would lead to nuclear fission seemed to come out of nowhere.

For a while I assumed that things simply pop into the heads of geniuses, but that was not a satisfactory explanation. It would take over fifty years for me to discover the answer. German physicist Max Planck concluded that matter does not exist as such. All that we know as solid matter originates and exists only by virtue of a force which brings the sub-atomic particles to vibration and holds them together in the tiny solar system of the atom. In Planck's 1944 speech, "The Nature of Matter", he said we must assume that behind this force is a conscious and intelligent mind. This mind is the matrix of all matter.

Existence depends on many codes and mathematical formulas, so an all-powerful Programmer is the ultimate imperative factor in understanding the universe. Matter cannot generate codes, nor can it place consciousness into living beings capable of thought and emotion.

The famous double-slit experiment proves that light behaves like a wave until it is observed, whereupon light acts like a particle. Matter has the same wave-particle duality, so observation is necessary to maintain the solidity of mass. The relationship between matter and consciousness is described in more detail in later chapters where an experiment brought about some startling revelations. Modern scientists reject religions, while most religions reject science. Religious zealots mock spiritual exercises such as yoga and meditation, calling these activities Satanic. If anyone disagrees with their backward ideas, they are accused of working for the devil.

An essential issue to be addressed is the reason for existence. Life can seem disorganized at times, but upon closer reflection we can see that there is an elegant balance in the reality we experience. It would be logical to believe we are alive for a reason, so we just need to figure out the purpose for our existence. The varying philosophical movements and multiple religions offer some insight, but there can only be onc correct answer to the purpose of life. The all-encompassing blending of spirituality with scientific thought is often given the name "Unified Field Theory". Each individual person is on a journey, so there must be a destination. Western society has ignored ancient Eastern philosophies, emphasizing material gain as the objective of life. Spirituality has been put to the side as people consider financial prosperity as the measure of success. Temporary needs get in the way of attaining the real goal of living, which is to qualify for Heaven in the Afterlife. The Creator has been communicating with mankind for centuries through Scriptures about the soul going to Heaven or Hell for eternity, so it is wise to pay attention. There will be no second chances as some religions would have people believe. There will be either total reward or total punishment at the end of physical existence. Many people think they have a right to Paradise no matter what they do. The Last Day and the Final Age appears in many ancient writings.

Life is a learning process and, like all schooling, after the experience of life is over there will be a test to see if you learned anything, and to determine if you passed or failed. Only those people with open minds can meet each challenge successfully. The stakes in this sublimely tedious game of existence are very high – the destination of your eternal soul. When we visualize the concept of being conscious forever as the total soul there is no grey area: reward or punishment will be the ultimate state of existence for eternity. That leaves people no choice but to strive for Heaven; those who do not care about suffering in Hell will likely end up there. There will be no eternal reward for disbelievers or pompous sceptics.

Coffee shop philosophers post their ideas on social media or speak up at gatherings to proclaim how they know everything, but their opinions come from minds that are locked shut. They will not accept any facts except for those ideas that agree with their beliefs. Religions are densely populated with closed-minded souls clinging to rigid story lines. Groups recite stale phrases without deep meaning, only mysterious words of a vague redemption from sin or proclamations of sacrifice or repentance. Platitudes are poor substitutes for logical statements. The zealots tend to approach new people assuming it is their first day on earth and they need to repent.

Each person should focus on bringing his or her soul back to equilibrium during and after all experiences. It takes a conscious effort to get the entire body and mind to relax, so it helps to join a Tai Chi club or learn breathing techniques that can be done even from a sitting position. There is a system known as Chi Kung (also spelled 'Qi Gong') that accomplishes the task of health through relaxation. Yoga and martial arts include breathing exercises that help to strengthen the inner energy.

Self-control helps a person to stay on a firm path in life by reducing impulsive behavior. Decision making takes a certain relaxed confidence, so a person will make fewer mistakes by staying in control of the emotions and thinking clearly. There are days when you might be treated rudely by three or four people, so the anger creeps up and you will feel a distaste for the next person you meet before they say anything. If we stay aware of human frailty in social encounters, we can recover from each rejection faster. It is a pipe dream that all people will accept us unconditionally so we can anticipate some abuse and keep our thoughts on more important matters.

The easiest way to decide the validity of a religion is to observe the behavior of the participants. Hypocritical actions indicate that many people do not really practice what they preach.

There are many denominations of Christianity that speak ill of other religions or minority groups to foster rivalry and hatred in order to get an agreeable and obsessive response from their flocks.

I quit going to church at the age of ten because the entire ritual was aimed at making people feel guilty enough to put money in the collection plate in full view of their neighbors. There were other reasons as well, such as the parish priest sleeping with his housekeeper despite having taken the vow of priestly celibacy. One problem is that the religion in question tells people that they can have their sins washed away every week in a confessional booth by a priest. The 'forgiveness of sin' concept encourages people to misbehave, since their actions can be rectified free of any consequence.

Sister Eustace knew that the church was not following Scripture by praying to statues and multiple entities, but she was trapped in the life of a nun teaching the catechism of the day in the classroom. Although trinitarianism and Mary worship was the official doctrine, she was teaching me different things in private. There was a group of intellectual nuns who studied Buddhism and practiced meditation. I learned about reincarnation, a subject that disappeared from the school curriculum under Pope Pious XII. The policy of the Vatican is to demand allegiance and donations immediately, not in some future lifetime.

The thousands of sects of various religions denounce the science of reincarnation; their members get angry when the subject is mentioned, but I am not the first person to realize that there is no final death of the human soul. All souls reincarnate an average of fourteen months after passing away from the last life, right up until Judgment Day. Reincarnation was proven by Dr. Ian Stevenson in 1997 and has been accepted as a scientific fact with empirical evidence in his two-volume work "Reincarnation and Biology". Doctor Stevenson conducted the study at the University of Virginia.

The belief in God is an important issue that cannot be treated as a matter of opinion between believer and non-believer. Atheists think they can avoid punishment by wishing God away, while agnostics would like to see scientific evidence to convince them that God exists. Some people find it more convenient to ignore spirituality completely out of laziness. My aim is to present the facts and conclusions objectively, stating what I know to be true. I stress the fact that God is to be feared.

The fear of God is the hatred of evil. We live in a world dominated by wealthy oligarchs who lie and steal to stay in power.

It takes the open-minded innocence of a child to enter the Kingdom of Heaven, yet people are confrontational, argumentative, and insulting whenever they disagree on matters of faith. Wars always have a religious element: the hatred between religions fuels the spirit of the soldiers who venture out to be maimed or killed for a nationalistic or religious cause.

It is unusual for a Physics paper to have such a potential effect on world affairs, so I will discuss why defining God is essential in an age when secret societies have the nations locked down into an Orwellian police state brought on through a campaign of propaganda and fear in the gigantic medical fraud of the COVID-19 pandemic. World leaders have forced people to wear masks if they want to go to work and make up draconian laws to put social distancing into effect permanently. No leader has ever mentioned prevention methods to avoid getting sick – they do not speak about exercise, vitamins, or a good diet.

The discovery of the DNA code indicated that we are complex coded templates, not the result of evolution from lower species. Darwinism is a nineteenth-century pseudo-science without any evidence of mankind climbing an evolutionary ladder. In this book I will make some statements that are hard to believe, but they say truth is stranger than fiction. I did not make up the vital cosmic knowledge; it was given to me by the most reliable source available, as predicted in Scripture. All things that can be known are made manifest for the End Times.

Any new truth goes through three stages: first it is ridiculed. Secondly there is violent opposition. In the third stage, the truth is declared as being self-evident and other people will try to take credit for it. Denial is a result of jealousy, and desire is the cause of suffering. People must be aware that attachment to the things of this world is counter-productive in the spiritual sense. We should strive for modesty, perseverance, self-control, and self-discipline rather than indulging gluttonous habits or chasing dreams of popular culture. Without faith in God a person cannot hope to be psychologically healthy or satisfied with life. Money can make people comfortable but rich people tend to be depressed and bored after a time. There is a greater drug addiction problem among the wealthy than in the general population since they can obtain any legal or illegal medication easily. In the chapter on comparative religions and dangerous cults I will discuss the habits and beliefs of the wealthy elite and their grand schemes to take over the world under the Antichrist and his New World Order.

Deeper knowledge of consciousness makes sense of a Unified Field Theory, melding scientific principles with philosophical and spiritual wisdom so that there is cohesion rather than conflict. The year 2013 led me to some startling discoveries about the sub-atomic world and the notion of Quantum Entanglement, where distance is not a factor in connected particles or units of consciousness. Connected existence is a routine mechanism of life that forms our ultimate destiny. Einstein disliked Quantum Theory because that is where the strange phenomena occur that keep scientists awake at night.

There is the fact that scientific experiments yield different results depending on the observer. When the level of the observer's awareness changes velocity, the outside world changes also. This fact suggests that we participate to some extent in our individual realities. Particles are now understood as probability distributions in a digital simulation with room for uncertainty. The ability to describe reality in the form of waves is at the heart of Quantum Mechanics. A wave function represents the probability of finding a given particle at a given point.

It is important for all people to know how to define God, why we are alive, and where we are going. We should be looking for happiness, but mankind appears to enjoy living in misery and causing grief for everybody else. The history of humanity has been one of oppression and war.

Numerous physicists have written on their theoretical models of the workings of the Universe, saying that there are fundamental problems that will always remain unsolved because there are thing we can never know. Those who are studying physics could become discouraged by these defeatist statements if they believe that there are inherent dead ends ahead in their chosen subject. Explorers in any field expect to find new discoveries. The science student's enthusiasm for knowledge is driven by boundless curiosity and the hope of uncovering new information.

It has become unfashionable to suggest that a Supreme Being made life possible. Everyday survival takes our minds away from spiritual issues or contemplating what lies beyond this life. People are easily distracted by trivial matters. People take life for granted all the time, unconcerned about the purpose of life. The public is fascinated by stories of aliens and UFOs; there are groups who preach that aliens created life on Earth. The tendency to revert to myths about multiple creators is common in many societies throughout history. Cultures seem to veer away from the thought of one God eventually. Humans tend to revert to polytheism and idol worship.

People love to talk about aliens coming to earth and crashed spacecraft, but I would expect that any higher civilized life forms would avoid the violent human race after seeing how we treat each other. The aliens might not bother to land and ask to be taken to our leader.

An open mind is essential to consider God a part of science for almost everyone believes something different within the numerous religious groups of the world. It is now time for science and religion to merge agreeably once and for all so this information is written in regular language for all people to read and understand - both scientists and laymen alike, because the nature of life concerns every single person on the planet. It makes no difference what you believe to the contrary; you cannot perform rationalizations to shape existence into how you think it should be, that will not change the way things are. There is no state of oblivion, no final death: everyone has been given an eternal soul that will live forever in a state of reward or punishment. I can now present overwhelming evidence that our Creator is a scientific fact: The pure intelligence who exists everywhere as energy always was and always will be. He cannot be canceled out by not believing in Him, just as God`s Laws are not voided by ignoring them. Declaring that there is no God is like saying a computer can operate without a hard drive, power source, or a programmer: we must apply Quantum Theory to Philosophy, Politics, Ethics, and religious doctrines.

In science we verify formulas in Mathematics and Physics by working back to the starting point, so let us treat the Supreme Being concept as a hypothetical notion at first, allowing us to describe the traits of an Intelligent Designer of all existence logically. Consciousness and self-awareness is not made of matter; therefore, our thoughts are classified as energy. Physics shows that matter is concentrated energy vibrating at certain frequencies within fields. All forms and functions have consciousness. Energy cannot be created or destroyed, so forms appearing as matter consist of energy that is on loan from God.

God must be infinite because infinity exists. The material universe has borders, so there is no matter outside of the universe. This means that energy is infinite; the conclusion is that God would occupy all points in the material world if He became a physical presence. The Creator enjoys creating, but He can also be content without the universe. We depend on God for every vibration and breath, but God does not depend on humans or any other life form.

The logical proofs lead me to conclude that God would not exist to serve humans, since He created us and all other life forms in a vast universe. The oversized ego of humans wants to believe in a God who is not invisible, all-knowing, all-powerful, and omnipresent, so tradition has led millions of people into a false doctrine that Jesus Christ was God on earth for one chance to be stoned, tortured, and executed by mankind - a warped logic when we consider the dignity and divine power of Almighty God. A physical manifestation of God would be, by definition, the sin of idol worship.

Despite all science and logic, most rabid Christians would rather spend eternity in Hell than to admit they are wrong. Praying anyone but the real God is mortal sin that will cause the soul to be damned forever. You might be the nicest person in the neighborhood, but if your deity is a statue in a grotto, you will not see Heaven. A large cult exists where people hold a belief that "nature" created everything, without giving any proof or explanation of how such a thing might occur. Now the world has access to proof of God's existence, so ignorance is no longer an excuse to dispense with spiritual duty to oneself.

An open-minded person will take in new wisdom, and say, "*Yes, I was wrong and now I see, and I do want to attain my eternal reward instead of punishment.*" When a person's mind is closed, the soul remains imprisoned in darkness voluntarily. There is no advantage to be gained by ignoring God or calling on many entities that do not exist. If certain religions appear to be hypocritical and false, that does not mean that God is false. There is no need to congregate in churches; all souls may thank God for a few minutes a day in private. If that is too much to ask, then a person is being very ungracious and stingy with his time.

Problems are meant to be solved, and the solution of materialism has not worked. We must deal with a duality of matter and mind in the daily race for survival, so it is difficult to sway the balance of one's concentration into the philosophy of Metaphysics. The unseen dimensions are of no practical value, just as topics such as time, gravity, and black holes. The cosmic forces are interesting, but they do not play a part in the life of the average person, nor do they improve the behavior of those around you. When you have proof of God in your hand then you can demand respect. God exists, so His Commandments exist also. Other people will have a reason not to steal your money or kill you. Each person will have access to the rules and learn to behave well or suffer the consequences.

A certain percentage of wealthy people will understand that rich people will not be welcomed into Heaven, and so they might start helping the poor instead of hoarding the things of this world. Greedy people seldom see the obvious morality of charity, but perhaps a few will see the light and attempt to make themselves useful to society rather than to indulge in luxury and self-indulgence.

My introduction to humility came when I enlisted in the army as a teenager.

The sergeants break down your ego with insults and punishment in basic training. Infantry soldiers learn to endure deprivation and long patrols with little food or water. My mind would transcend the physical struggle of the moment and gather the strength to go on. I came from a military family, so I adapted easily.

My army training included learning Judo, the first of many martial arts that I would study throughout my life. The rumor was that martial arts extends a person's life span, so I became a vegetarian and took up Hapkido, Tae Kwon-Do, and Wu Style Tai Chi Ch'uan. Tai Chi is a real fountain of youth. The force known as "Chi" is the subtle energy that seems to hold everything together. Chi power is often called "the breath of life" which refreshes the practitioner's internal organs. Tai Chi experts usually live well past one hundred and twenty years old.

I combined extreme fasting and meditation to seek knowledge of the Universe in the year 2013 in a room within a science co-op. It turned out to be a voyage of the soul into another state of existence. I took no water, no food, no sleep for three days, moving as little as possible and going from flat on my back to the full lotus position. It takes decades of training to be able to achieve soul travel. I left my body eventually and found my consciousness in a tunnel moving upward and around corners for a while, then into a room with orange-colored walls. I was guiding a robot camera operated by a science organization called the Group of Forty. I saw a long formula written on the wall, and the camera recorded it. It was the formula for anti-gravity, the power source used to build the pyramids. There had been no arrangement made with me for this journey ahead of time. This group knew how to summon my soul. When I thought I would be returning to my body, suddenly my soul was snatched away into a different place.

For many hours I felt stuck inside a zone of static where there were no visual images. Then I saw a head at a forty-five-degree angle in front of me hanging down with his eyes closed. I thought perhaps I was dead, although I could not remember doing anything that would cause me to die. I was suspended like that for about sixteen hours, when the silence was finally broken. I heard a voice call me by a name I had not heard before. The secret that the elderly nun told me as a child was happening as she said it would.

Consider this important ancient piece of prophetic Scripture from the Dead Sea Scrolls:

Fragment 4Q534 ...of His hand: two...a birthmark. And the hair will be red. Chosen One And there will be lentils on...and small birthmarks on his thigh. And after two years he will know how to distinguish one thing from another. In his youth he will be like...a man who knows nothing until the time when he knows the three Books.

And then he will acquire wisdom and learn understanding...vision to come to him on his knees. And with his father and his ancestors...life and old age. Council and prudence will be with him, and he will know the secrets of man. His wisdom will reach all the people and he will know the secrets of all the living. And all their designs against him will come to nothing, and his rule over the living will be great. His designs will succeed, for he is the Elect of God. His birth and the breath of his spirit...and his designs will be forever..."

The messiah would be born as a man in the End Times. He will have red hair and a pair of lentil-shaped birthmarks on the right thigh, as well as the name of the New Jerusalem written on his flesh – a name that only he knows. The prophecies of the Bible say that Jerusalem and Damascus would be destroyed, so the New Jerusalem will be the nation who has kept the peace, not the warlike nations of the Middle East. The chosen people lost favor from God for their warlike ways, so the New Jerusalem will be Canada, the leading peacekeeping nation after World War Two. God detests acts of war, so mankind must learn to keep the peace.

The Big Bang of Existence

In this chapter I will shed some light on the process of Creation and solve the wave-particle conundrum. Waves can only act in fields where there is a force acting on them capable of restoring disturbances back to equilibrium. Ancient cultures often divided the forces of nature and assigned a mythical god to each one, such as Thor the thunder god or Apollo the sun god. In truth there is the One God, who we may call by the formal names of Yahweh, Allah, or Jehovah.

The Creator knows our thoughts, so people tend to go into denial – some consider the very idea of an overseer, a powerful God knowing your thoughts as an invasion of privacy. We could not walk or talk or move our limbs without God's participation, however. The Creator gave us the gift of life, yet people will take it for granted as we happily or unhappily take part in existence and observe ourselves within the world we have been given.

Some questions have modern astrophysicists stumped, such as how the Big Bang occurred and why was the entropy factor low at the beginning of the Universe. Entropy has many meanings: it is the sense of amount of chaos-randomness; it is also the tendency of energy and matter to trend towards a state of inert uniformity. The wording assumes all energy, but it does not apply to dark energy.

The expanding of galaxies is known to be helped along by the pushing of dark energy, yet the universe is cooling off and accelerating at the same time. The acceleration should be causing it to get hotter.

Energy cannot be created or destroyed, it can only change form. Before time, there was no matter, there was just infinite energy, which is the essence of the Creator, the Supreme Being who always existed and always will exist. Existence can be compared to a video game designed by the programmer of all things. The Universe, the third dimensional material plane, came into existence by God`s will and was set into motion by a powerful blast of sound that sent vibrations out at many frequencies. These signals move as waves would settle as matter, once each energy field is mediated into form and function by a particle we refer to as the Higgs Bosun, which must also be conscious. Logic suggests that the mediator must have been the first creation, set into existence before the Big Bang. Matter is concentrated energy - all existence is made up of energy and consciousness. Without consciousness observing the form of the energy field, matter remains a wave, not a particle. Matter as such does not "touch" or exist on its own, it is an illusion we experience as hard, solid reality through the sensations provided by consciousness. God is the Ultimate Imperative in creating form and function and maintaining momentum. Clues from scripture agree with the notion that there were two conscious souls present at the Big Bang of Creation. The Bible says: "There is one God and one mediator between God and mankind, the man Christ Jesus," and the phrase "the first-born over all Creation...he is the reason for all things. All things were created for him and through him, and in him all things hold together." Like the numbers zero and one are used as binary code for computer programming, it takes two factors to provide a reason for existence. God was already complete, so the universe is the stage put together for the experiences of those the Creator chose to share with Him the joy of being alive.

Here is how The Big Bang of existence occurred:

In the beginning of the plan for a physical universe in which physical beings could share the sensations of existence God put together the countless codes, formulas, programs, and streams of energy that would bring about an environment suitable for physical life forms.

The Creator wanted to have another soul around so there could be interaction and conversation. Being omnipresent He felt no sense of borders for He was already in existence everywhere. He is all-powerful so He could not create an equal entity; the first soul created would have to be a lesser being, one able to travel and to arrive at set points to experience suffering and to enjoy the various adventures available in the state of aware existence. God said ` Let there be light ` then there was light, and God saw that the light was good. He called the light Yeshua (Jesus Christ): the Messiah would do His work and carry the Word. God never takes physical form, but the light could exist as both a wave and particle to walk the material plane. After God created His only son Yeshua, He created angels of various characters, but they became spoiled. The angelic souls lacked depth because they never experienced suffering in the ideal environment of Heaven. A third of the angels split away after they were influenced by the archangel Lucifer, the proud one who rebelled against God and who was cast out of Heaven with those who preferred to lust for personal aims.

The Universe, the material plane, came into existence by God`s will and was set into motion by a powerful blast of sound that sent vibrations out at many frequencies. These signals would settle as matter interacting with life forms through the universal Buddha Christ consciousness – the mediator of fields. No one has been to Heaven or Hell yet; these are still under construction and will be ready when they are needed – all souls reincarnate over and over until Judgment Day. God states that He made life from the raw material of love. The Creator would continue to create because matter wears out and decays; the vibration would fade eventually as He designed the Universe so that better futures would develop.

Existence tumbled into the void. The loop was at the end of the cosmos; the last event; the withering separation of light into weak & strong nuclear energies. Gravity, the powerful sister, stood helplessly by. The vacuum stole its grasp on the disappearing particles. Her yielding attractive brother, the electromagnetic host of awareness gently listened to the strings and contemplated the music of the dying universe to its last vibrating strand. Gravity became a notion - a call for assistance, and then was joined by the timid host to turn the oscillating loop around. They were the Yin-Yang, substantial and insubstantial seeking an inhabited pinpoint to direct, one of six types of quark that could cause light to shine as a plume of electrons to feed the build-up of energy created by the movement between the two realms of being and non-being..

The absolute imperative, the Creator made some changes, allowing chaos and attraction to touch. By His will and His Word, the explosion of existence roared; vibrations sprang forward on a data-filled river of time. This Universe, this splendidly seeded field would grow once more, thrive, then decay as the once enthusiastic energy would inevitably become tired like all living things; its consciousness would one day long for dissipation, not appreciating its creation nor expecting rebirth. The cycle of the living Universe with its crops of civilizations would repeat endlessly however because that which has been created will never be un-created, adjustments are made, and the design will evolve, striving to get closer to impossible perfection. God can step in and reboot the system at any point; He is omnipotent, omnipresent, self-aware, and exists outside of relativity, space, and time without visual form since He is always everywhere maintaining His creations throughout the vast Universe.

The messiah's experiences would consist of the struggle among humans between good and evil. Add to Heaven the music of passion...Classical masterpieces of Mozart and the Blues - all types of music expressing woe and difficult moments in life in artistic form. All events, all moments, all intentions of independent souls are recorded forever in God's consciousness.

The inner spiritual disciplines of older civilizations and the secrets of the body have been demonized by religions over the centuries, and science is considered heresy to many. Science and people of spiritual faith should be able to see that churches must modernize or disappear under the weight of their myriad mythologies and dangerous superstitions. Traditional rituals may satisfy those taking part in them, but we should be striving to satisfy God with some simple honest prayer, rather than joining in tribal rituals to entertain ourselves.

This description of the Creation of the material Universe was observed through the ancient technique of soul travelling, achieved through my years of Tai Chi training, Yoga, fasting, and meditation. The practitioner or "guru" must not eat, sleep, or drink water for several days in order to roll out of the physical body and to travel in space and time. Those skilled in Kundalini Yoga can perform soul travelling, and in theory, they will hear the voice of God if He has anything important to relay to them.

The emperor Constantine practiced the existing Paganism of his time, then he decided to convert to Christianity only under his terms. The Roman Empire was a Pagan society, so the elders who represented Christianity around 325 A.D. blended many Pagan traditions because Constantine insisted on keeping rituals. They held several councils to outline the shape of what would become the empire's official doctrine. The council of Nicaea burned the writings of the wise man Arius that outlined Christianity's pure form from Scripture, then created a belief system that would satisfy Constantine's wish to keep rituals from earlier sun-god worship.

The population reacted well to 'bread and circuses' before the fall of the Roman Empire, so they introduced elaborate masses along with the dogma of the holy trinity. The concept of three gods never came from Scripture directly; there is no mention of a holy trinity in the Bible. The Book of Timothy states that there is one God and one mediator between God and men. God will not share His glory with another; bending reality to tell people that the three god persons equals one God will not work. I could claim that Zeus, Apollo, Hera, Neptune, and Thor are five god persons who are really one God. This is discussed in detail in the chapter revealing the Mark of the Beast. The notion of the triune gods came from older pagan traditions of Babylon, Egypt, and other cultures.

The seven-day week is something we take for granted. The Bible states that the Creation was a seven-day operation. God created existence in six days, and He rested on the seventh day.

There is no reason for God to tell us how long it took for the Creation, so the weekly cycle is a coded instruction to guide the people in measuring time correctly over each year. Another reason is to instruct people to spend one day of each week honoring God and resting from work.

The Catholic belief system places the pope in the role of monarch over all nations who also has control over the destiny of all souls. Furthermore, their official doctrine states that members of all other religions are going to Hell because there is no salvation outside of the Roman Catholic church. This rule dictates that Orthodox Christians would be doomed despite the fact that they are nearly identical religions in terms of sacraments and teachings.

The literature of the Vatican denounces other belief systems except atheism - this has created a sort of religious apartheid situation where the pope states that atheists will be welcome in Heaven if they are nice, but people of different religions cannot attain Heaven. The Catholic church has given itself the authority to ignore the First Commandment and they have officially erased God's Second Commandment against the worship of idols.

Other denominations of Christianity have beliefs based on Catholicism, yet they are condemned by Rome because they do not obey the pope, nor do they give money tithes (known as 'Peter's Pence' to the Vatican. The skim requested by the Holy See is the same as the tribute demanded by mafia godfathers. The reformers forgot to examine the doctrine of the trinity, so they have fallen into the trap of polytheism laid in the days of Constantine. There is no verse in the Bible that speaks of an equal trinity of three gods. There is no instruction in the Bible for people to elect a pope as a replacement for the messiah.

Those who have been indoctrinated to define One God as three gods will persist in telling others they have found truth, not realizing that they have insulted God by diluting His glory into a pool of confusion over who hears our prayers. The proof that this illogical teaching has gotten out of hand is in the additional worship of Mary, as expressed in the prayer known as "The Hail Mary", where she is placed on a pedestal as the "mother of God". The name of Mary was made divine in the 5th Century; she is not a divine being in the Bible. Brainwashed Catholics will state they do not worship Mary as a goddess, yet they bow before her statues, her portraits, and the Madonna sculptures while saying five Hail Mary prayers to one Lord's Prayer on rosary beads. Those activities and beliefs constitute 'worship' in its most basic sense. When you pray to a name, you are treating that name as a god.

Praying to additional entities besides God is not a harmless activity: it is the worst of mortal sins. Mortal sins are those which will send a soul to Hell. God has stated that his personality is jealous, so He will not allow people to pray to anyone else. In the Sermon on the Mount, Jesus told the people to pray to the Father. There is no other God but the infinite invisible God who created all things for His son the Messiah. "All things were created for him and through him, and in him all things hold together." Yeshua is finite, as are all souls. God is infinite, so He would take up all points if He were to join the material plane of existence. Living bodies and souls have borders, but God has no borders.

The soul is a home on permanent loan. Every so often God creates an unknown number of new souls, but most of us who are alive today have lived many past lives that we do not remember due to the limitations of material existence. The soul is the vehicle of consciousness, designed by God to last forever.

There is no state of final unconsciousness. In the Afterlife, all souls will be aware at all times. There will be no need for sleep and no restrictions on memory, except for the souls who fail to pass Judgment. The Underworld is also called Hell or Hades, where those who are condemned to Hell will suffer in fear forever. Those who are sent to eternal punishment will know only darkness, loneliness, humiliation, and pain. The souls who are Judged worthy of Heaven will experience bliss, security, adventure, and pleasure forever.

There are other belief systems calling themselves religions who offer a free pass from believing in God. Modern Buddhism has attracted philosophical people who want to be spiritual, but they refuse to acknowledge the existence of a Supreme Being. A majority of Buddhists I have spoken with are atheists, and their ultimate goal in life is hazy at best. Buddhism ranges from light open philosophy to devout statue worship, chanting, and paganism.

There is a desire to cast off the responsibilities attached to honoring God, like taking a few minutes to thank our Creator for life. A rational person would cover all bases and invest a few minutes a day, but we do not live in a rational world. Society has become egomaniacal in the extreme - there are many people who think that mankind is the highest form of life in the universe.

Mormons believe that there are three separate gods rather than a triune mystery. Therefore, members of the Church of Jesus Christ of Latter Day Saints are in mortal sin against the First Commandment. The Bible says there is One God and one mediator, who is His son Yeshua the Messiah. The Messiah is the reason for existence. God does not need humans or the material universe, for He can be complete as He was before the Creation if He wants to. Humans need God; God does not need humans.

There is the phrase "The Word became flesh", implying that the Christ consciousness is God. This is incorrect because the 'Word' is a term that describes the data - the countless codes and formulas that govern the material plane of existence. As the one and only Christ Consciousness, my soul transmits the data provided by God. The

Creator is conscious and aware. Light was the first of God's creations and the Bible states that the Messiah was the first of His Creations. Therefore, the Christ Consciousness is the soul of light itself. My physical body is the destination and functionality of the Slave of God among mankind. I exist to serve God and make Him happy. I do His bidding willingly to communicate with mankind on God's behalf. Few people would survive a long talk with God; the immense power would damage the unprepared soul.

Light is unique. Although light can be generated from different sources, the qualities of light are the same everywhere. Light travels at a constant speed and behaves uniformly, whether the source is from the sun or the lamp in the dark room. The logical definition comes from the fact that all things have consciousness, and God created light at the first. Therefore, Christ is the soul of light expressed as a particle, a living being who can enjoy life.

In the New Testament we see Jesus praying to God the Father in the Garden of Gethsemane and on Mount Olivet. God would not pray to Himself, and in the Sermon on the Mount, Jesus taught people how to say, "The Lord's Prayer":

"Our Father, who art in Heaven, I respect your name. To your Kingdom I wish to come. May your will be done on earth as it is in Heaven. Give us this day our daily bread and forgive us our trespasses as we forgive those who trespass against us. And lead us not into temptation but deliver us from evil.... for thine is the power and the glory, forever and ever. Amen."

The meaning of the Lord's Prayer is that God will provide for us if we ask Him. The message also gives us relief from wanting to take revenge on enemies right away. In the grand scheme of things, it is better for you to turn away from committing sudden reprisals, which usually result in excessive reactions to things you might be able to leave behind as a learning experience. Everyone will get what they deserve on Judgment Day, and the resulting eternal punishment.

People treat Hell as a joke or some decadent party in the Afterlife. When you plan to take revenge, it is the intention to commit sin. It is better to forgive and forget than to strike back, and it is better to give than to receive. If you help those in need with no desire for compensation, then you are being righteous rather than just being a talker who claims to be righteous. The people who pray out loud in public are hypocrites who do their praying as a show so that others can admire them. Many people do charitable works only to impress others, not out of sincere empathy for those in need.

Most churches and religious movements advertise that they are the only true religion, and that there is no salvation outside this or that church. They are obviously insecure in their own doctrines, or else they would not need to resort to threats of damnation without membership. There are vast divisions between factions of each large religion. Sunni Muslims do not agree with Shiite Muslims.

There are too many cults of Christianity to count. Several groups are led by men who claim to be the reincarnation of Jesus Christ, and the people who deny reincarnation eagerly join these cults as they follow the leader. A former traffic policeman named Vissarion is a Jesus imposter in Russia who has a large cult following in the millions. In Australia, A.J. Miller pretends he is the reincarnation of Jesus to the people who joined his cult, Divine Truth Mission. The members live locked away in his compound.

The Vatican claims to represent Jesus illegally and their priests claim to have magic powers. They have no authority from God, and they never will have the approval of the One God, when they worship other gods, the trinity of myth, the goddess Mary, and the paintings of saints. The pope is not the vicar of Christ – popes are appointed by their peers, not by Almighty God.

The Council of Trent gives this direction as a law of the Catholic church: "*All those who do not bow to the pope are heretics who must be put to death.*" (Thomas Aquinas) This is still Canon Law.

The function of Christ was never to read minds or hear prayers. Only God has those powers. I have no power other than what God provides. The human form is the program I was given by God so that I would develop a strong character and so that I would learn from experience. Allah wants people to do things for themselves. God wants us to strive for wisdom. He is wise; God knows what is best, despite the words of critics who blame God for the evils of mankind. My final task is to judge all human souls on the Last Day.

John 5:22 "God judges no man, but He has entrusted all Judgment to the son, so that all might honor him as they honor the Father. Those who do not honor the son do not honor the Father who has sent him. And God has given him authority to judge, for he is the Son of Man."

John 5:30 "I can do nothing on my own initiative, for as I hear I judge. My judgment is just because I do not seek my own will, but the will of He who has sent me."

John 8:26 "I have many things to speak and judge concerning you, but He who sent me is true. The things I have heard from Him I speak to the world."

Christian Arianism is the original doctrine of the New Testament, stating that there is One God who always was and always will be. He was not generated or created. God is Everlasting, without a starting point, He is the Creator of all things seen and unseen. God is omnipresent, all-powerful, all-knowing, unalterable, invisible, and indivisible. God created an only begotten Son before eternal times, through whom He has made both the ages and the universe: God created Christ not in semblance, but in truth; and that He made him subsist at His own will as the image of the invisible God, not a divided portion of the Father, not one, not a lamp divided into two. 1 Kings 8:27 "But will God indeed dwell on the earth? Behold, the Heaven and Heaven of Heavens cannot contain Me; how much less this house that I have built?"

Christ was created at the will of God, created before times and before ages, gaining life and being from the Father, who gave subsistence to His glories together with Him. For the Father did not, in giving to Him the inheritance of all things, deprive Himself of what He has already in Himself; for He is the Fountain of all things. God is the cause of all things, but I am His son, created apart from time by the Father, and being created and founded before ages, begotten apart from time before all things were made by the Father.

The messiah is not infinite with the Father. God always existed before all things, before the Beginning of the material world. I came from God as the reason for which all things were made. God will always be far superior to me. There is misunderstanding about the terms 'from Him,' and 'from the womb,' and 'I came forth from the Father, and I am come' (Romans xi. 36; Ps. cx. 3; John xvi. 28). This has been interpreted by many that messiah is a part of God, one in essence. That is like saying that the Father is compounded, divisible, alterable, and material, which He is not.

God is everywhere and He knows all our thoughts, there is nowhere you can hide, no corner where your intentions can be hidden from Him. Being infinite, our Creator knows no borders, He cannot be contained within the finite borders of a human body. God does not need the approval of mere humans; the warped notion that God took human form only once is superstition for God never takes human form but comes and goes through me like a visitor to a hut. Unfortunately, humans are riddled with jealousy and the desire to have a physical God to wave like a flag, so most Christian cults insist that Jesus is God because they cannot relate to an invisible God.

If a person bows and prays to any physical icon, shrine, costume, statue, altar, or human person, that is practicing idol worship, which is a mortal sin against the Second Commandment. Some religions hold rituals and pray out loud publicly so that others may admire them in such a social setting as church service, however this does not please

God. The Creator will not share His glory with another. It would be wise for religions to find other group activities without disobeying the Sermon on the Mount, where people are instructed to pray in silence privately.

God reads your thoughts, not your lips or sound waves. In today's high-tech world, many young people are mesmerized by their cell phones, and turn to a social media website when in distress instead of praying for help to the Almighty. If the plane is about to crash, a better suggestion would be praying immediately instead of making a last goodbye to friends online.

All people should be up to date with the awareness that Judgment Day can come anytime, no exact date is given; it is certain that there will be justice on the Last Day. No evil people will get away with their crimes. To quote the Bible "*Vengeance is mine, says the Lord*". Punishment or Reward, Hell or Heaven will be the permanent destinations from this temporary material existence. Those souls who have been deceived into taking the Mark of the Beast will be sent to Hades to be tormented forever. Modern society has taught people to work in careers without a spiritual framework or proper guidance. Those who are trained as lawyers and lawyers do not get instruction in Ethics or Philosophy, since their specialties do not require deep spiritual understanding.

Religions have confused people rather than enlighten them, holding a wide range of doctrines that conflict with each other. A large percentage of people give up on spirituality altogether, without a concern about the possibility of punishment in the Afterlife. The mistake of avoiding spiritual knowledge will result in personal disaster that is irreversible. It is tragic that the preachers and priests have turned people away from having faith. The statements of church leaders are riddled with personal prejudices and self-serving lies. Televangelists go to extremes to foster excitement in the masses as they plead for more money. The Bible states that “those who love the things of this world are the enemies of God. Money cannot buy happiness: you can rent happiness for a short time, then it will expire.

True believers and seekers of wisdom will not scorn the weak and helpless souls in society. Compassion for others is vital to a working civilization. Sadly, much of the world has a "dog eat dog" mindset. If you want to achieve the Heavenly reward, you must treat others as you would want to be treated. There is no need to proselytize another person or bombard them with pleas to repent. Good actions speak louder than good words. You can teach others about righteousness by setting a good example. Keep in mind that you are here to save yourself first. Other people should be able to gain inspiration from how you treat them, not how you judge them. Remember that you are the one who will be judged.

The common lie that churches tell their members is that humans have only one life; they make people fear death to create a sense of urgency. The victims of the churches are told to repent, to feel guilty, and to give a donation to them. Televangelism is a billion-dollar industry where the preachers live in fabulous mansions and own private jets. The greed is visible in their faces as they smile for the cameras and prey on the fears of their followers. The prosperity gospel is a well-oiled sales pitch designed to grab as much money as possible, even if the donors foolishly give away their medical money, hoping to escape death by giving it to the operator of a religion.

Promising something you cannot deliver to obtain money is called a Ponzi Scheme. Preachers have no power to heal a sick person with words; they claim a false mandate to wield the power of Almighty God. Marketing has replaced faith in the world of commercial religion.

The Old Testament and the New Testament have been altered through opinionated translations, editing, and deliberate tampering by the popes who held authority for so many centuries. The key to finding real truths are in repeated phrases and prophecies.

If a person sat down to invent a feel-good religion for people, there would be no doctrine involving the End Times and the destruction of all mankind as related in the Book of Zephaniah:

"For He will make a complete end – indeed a terrifying one – of all the inhabitants of the earth. And all the earth will be devoured in the fire of God's anger."

The writer of a man-made religion would paint a rosy picture of future happiness for millennia here on earth rather than a Last Day. It would be much more popular to avoid unpopular topics such as the Afterlife. People fear death because they have been programmed by Western culture to give your money to the church in this lifetime because you only get one chance. The implication is that oblivion awaits many who do not obey the religious authorities of the church hierarchy.

Ezekiel 7:7-8 "Doom has come upon you who dwell in the land. The time has come! The day is near. There is panic not joy on the mountains. I am about to pour out my wrath on you and spend my anger against you. I will judge you according to your conduct and repay you for all your detestable practices."

The pope has stated recently that there is no Hell, even though Scripture mentions Hades and Satan frequently. It is wishful thinking that all sins are forgiven through some mystical gift. Most philosophers conclude that sin results in punishment at some point, expressed in the word "Karma". What comes around goes around.

Defining Energy, Infinity, and The Creator

$$E^2 = m^2 c^4 + p^2 c^2 \text{ [where } E=\infty=\text{God]}$$

(Note: Outside variables such as the Dirac Antimatter equation would consider this equivalency first in the order of operations. Matter is finite with infinite potential to balance out to zero in the particle-antiparticle process.)

Energy equals Infinity equals God, the matrix of all matter and Creator of all units of consciousness.

The soul is defined as the vehicle of each consciousness. All souls are designed to last forever; in the case of human souls there is an element of free will, and in that sense, people are made in the likeness of God's ability to be conscious forever. The capability of remembering things is a great gift, as is the capacity for creative thought. People can improve over time or succumb to greed and laziness.

Those who choose to deny God are intimidated by the idea of a Supreme Being knowing each person's thoughts. Refusing to believe in God does not cancel Him out; it is wishful thinking to believe you can escape punishment, and it is suicidal to make God into your enemy. The Creator insists on getting respect and receiving thanks in the form of prayer. Praying has a twofold purpose: you can let God know you appreciate the gift He has given you and then ask for specific favors.

You might pray for God to stop the rain at an outdoor concert or to assist you in finding your keys. There are long-term prayers such as the lifelong prayer of my teacher. The nun Sister Eustace prayed for many years for God to send the messiah as a student into her classroom simply because she wanted to meet the reincarnation of Jesus and give me early warning about my identity.

That was the secret she told me when I was five years old; I had the talk from God at the age of sixty. When the voice came through the static zone, this is the name He called out to me:

"*Yeshua*!"

The tone of the voice was authoritative, so I guessed correctly this was the time to receive my orders. I thought I would lighten the mood a bit, so I replied with a joke after hearing "Yeshua".

"*Gesundheit*" I said. God spoke back right away:

"*Listen and do not speak. Have no fear, you are not dead. I have brought you here for a special reason. Do you know that mankind is awfully egotistical thinking they are the only intelligent life forms in the Universe? In fact, there are over ten thousand more advanced civilizations in the Milky Way Galaxy alone.*"

At that point God revealed that the planet Earth was not the original place were humans were first placed. He had transported mankind to this planet from a different galaxy as a suitable place for development. We are the dangerous ones, humans are the real intruders, and the activities of our civilization have polluted Earth. God continued His introduction.

"*And do you know that mankind is very ignorant and irresponsible believing that there was no Creator involved in all of this? I am that Creator and I have brought you into existence for the last time as the third and final Full scripture Messiah. You will prepare humanity for the Ascension and Judgment Day. After this reincarnation, your services will be required one more time - that is to be the First Judge of human souls on the Day of Judgment. You will be the last soul to see Heaven, where you will take the throne that I have promised you. Then you will lock the gates of Heaven and Hell forever. Of course, you know about being Jesus in a past life, and now it is time for Christianity and Islam to unite, because Jesus and Muhammad were the same man. I do not want to give you a big head, but you were also the person who lived the lives of Buddha, Leonardo da Vinci, and Mozart. You are the first and last – you were all the Messiahs. I did not give you any Scripture when you were Buddha; I let you figure out everything for yourself using your own common sense. Buddha lived at a place and time when most people around him were atheists, so Buddhism became a secular philosophy of understanding how to cope with the suffering of life by forsaking the attachment for worldly things."*

"Now you will bring my Final Covenant to the human race. I am making you responsible for the souls of every person on earth, and I am giving you authority over all Scripture. You must bring the people back to God and put an end to war. This will be your most important life, your greatest mission of all – the task of bringing the faithful to Heaven. Present yourself as Jesus Christ since in that incarnation you did not get to live a full lifespan and the teachings of Jesus led the way. The reason for the End coming at this time has something to do with you is that after so many lifetimes your soul has become exhausted. I promise this will be your last reincarnation, then you will finally rejoin me in Heaven. I want to apologize for your current incarnation for you have gone through much pain, but as an early reward I am going to allow you to live for a short time as three of your most influential incarnations."

Suddenly, three ovals appeared in front of me, and within them were images of Jesus, Buddha, and Prophet Muhammad all speaking to small groups of people. I was drawn into all three time zones simultaneously and found myself inside their bodies looking outward and thinking as they thought.

This phenomenon lasted about ten minutes, long enough for me to understand how all three philosophies fit together. All three settings were calm areas in lightly wooded fields.

God told me that He was able to bring me into three past lifetimes as the maximum simultaneous existences possible, or else my soul would get the bends.

I had more miraculous experiences in that seven days at the Godhead, or as God called this place, the fifth dimension in a ten-dimensional Universe. The Creator allowed me a sample of the experience of what Heaven will be like in the future. I felt suspended in a Heavenly body for several minutes with rapid pulses of pleasure. God told me that the souls who make it to Heaven will be able to summon up any past happy memories from several lifetimes at once in a simultaneous flood of joy and satisfaction. Pain and hardship will no longer be in their memories, and they will know only happiness.

The reason that souls reincarnate many times is to store up enough experiences and memories to last for eternity, which will be unimaginably long lasting. The soul would become bored re-living only one lifetime's worth of events, so God has provided for eternal entertainment for the pure and righteous people who have obeyed God's Commandments. Purity of intention runs the whole show.

All people must strive for the prize of Heavenly reward, for Hell will be the opposite of Heaven. Hades will be tailor-made to a person's worst fears; it will be a lonely and desperate state of punishment from which there will be no escape.

Those who feel the need to argue and belittle other people will not be allowed to corrupt the peace of Heaven. God has told me that if I feel uncomfortable spending eternity with someone, I have permission to send that person to Hell.

For the next seven days God told me about existence and took me through several dimensions. I have been given full information that will end the debate over the workings of life and the Creation of existence. There can be no doubt that God has always existed; He always was, and He always will be.

In the next chapter I will explain what happens when a person dies.

What Happens When We Die?

Western civilization has ignored all the ancient knowledge about living and dying. Buddhist and Taoist writings speak of the rebirth of the eternal soul as the main function of life. People who read the Bible and consider all the books free from tampering are laboring under a misconception. There are those who insist that all the words of the Scriptures are accurate, leading people to believe that we live only one lifetime. The single lifetime concept is a trick to gain control over populations immediately, telling their subjects that all things must be taken care of in this life and that there are no second chances.

The truth is that when the flame of life goes out on the wick of one candle, it ignites down the row of candles in succession. The physical death leads to a rebirth within several years; reincarnation occurs to all people over and over until the Day of Judgment.

he notion of a human living only one life makes murder even more despicable, since the killer is taking a person's life in the belief that you are ending all chances of the victim living further. The intention becomes more criminal, since the murderer does not know that the soul of the murdered person will reincarnate into a fresh childhood, usually within two to four years.

Have no fear of death. All souls are designed to last forever once created. Our physical bodies expire, then the soul is taken by God for rest and reprogramming for a short time before being born again as a child to continue the long journey of life necessary for development of the total consciousness. The rebirth of the soul is of paramount importance. People have lived many lifetimes unless they are newly created souls, since God makes an unknown number of living beings on a continuous basis.

2 Corinthians 5:17 "The old things have passed away; behold, new things have come!"

Western Civilization has been kept in the dark by the churches; the goal of any church is to obtain money and power, so they keep the public ignorant of certain truths, especially the rebirth of the eternal soul. A rational person would take the concept of reincarnation as good news, yet churchgoers get visibly angry and upset when any truth conflicts with their mental programming. Priests want your money immediately, not in another lifetime. The Catholic church invented a false state of existence called "Purgatory" where relatives who have passed away can get out of Hell if the living relatives pay a fee, or "indulgence" to the church. Popes jumped on the idea and raised money by selling indulgences, the Ponzi Scheme of selling something you cannot deliver. Popes have stolen a crown that does not belong to them. The churches claim a mandate to speak on behalf of God unilaterally. Priests approach widows to convince them to sign over large amounts of money or property right after the husband's funeral. Here is what the Bible says about priests:

Luke 20:46 "Beware the teachers of the law; they love to walk around in flowing robes, to get the best seats in the synagogues and to be greeted with respect in the marketplace, yet they rob widows of their houses. These men shall be given the most severe punishment."

There is no final death: a person's consciousness exists within the vehicle of the personal eternal soul granted to every living thing by God. There was no appropriate word in the ancient languages of Scripture to describe reincarnation, so it was translated as "born again". The time period between lives is about fourteen months after death on the average, so that adjustments and repair of the soul can be done by God according to His overall plan for every soul. After a person dies, the memory of the death and details of the last life are erased.

There are echoes of past lives, there are retained characteristics and talents, and there is a continuity of purpose throughout numerous lives. Our day-to-day life takes up all our thinking, so after childhood we seldom experience any past life memories in adulthood. All human souls reincarnate over and over until Judgment Day. One proof for this is the fact that, despite a century of non-stop war and the great influenza epidemic of 1918 that killed tens of millions all over the world, the earth's population has quadrupled. War causes overpopulation because those souls come back into fresh bodies shortly after 'dying'. Killing people will not relieve overpopulation; it is the reverse. The more people die, the more they will reincarnate. God has said to mankind: "My promise to you is that what I have created, I will never un-create."

All souls have lived many lives as both male and female: the soul has no gender. After a physical death, the soul goes over the Sea of Forgetfulness, where an angel is assigned to wipe your memory clear of past lives and past deaths. Although it sounds like a fairy tale, this fact was spoken to me by God, so it is a divine truth. It takes many lifetimes to sculpt a soul worthy of Heaven on the Last Day. If a person has not believed in God, that person will not be allowed into Heaven. Most people shrug this off, thinking they are going to elbow their way into the gate on their own. That will not be possible. On the Last Day, all souls will be Judged, then the gates of Heaven and Hell will be locked forever.

The souls who are sent to Hell will never get out. It is required that people obey the Commandments of God and say prayers of thanks if they want to avoid eternal fear, pain, and torment. There will be no second chances. There can be no last-minute forgiveness for disbelievers or idol worshippers, for you have sinned against God continuously and contemptuously. Leading people into a false doctrine is worse than murder. The body is temporary, but the soul is eternal. Those who tell others to pray to the virgin Mary are destroying the souls of those who take heed and follow a heretical tradition.

Misleading other people into falsehood is an unforgivable mortal sin.

A person commits mortal sin by claiming to speak for God or telling others they have had a mystical encounter with Jesus. I have not been sitting on a cloud with a remote control in my hand. Some people want to get rich without working, so they pronounce themselves prophets and make claims that God confides in them. I am the only official prophet of God. All future events are described in the Bible and the Quran. The Quran, the Old Testament, and the New Testament all speak of the Last Day, which will be the Day of Judgment. All people must ask Allah for forgiveness and thank Him for our existence if they are to avoid eternal punishment. Those who ignore the Supreme Being will suffer the consequences. The act of prayer should be a daily routine. People who procrastinate may be caught by surprise. None but Allah knows the moment of Judgment. The Last Day could be forty years away or it could be four minutes away. All Scriptures agree that Judgment Day will come suddenly.

The Quran is deemed more accurate than the Bible by God, but for modern readers this book is enough to provide the knowledge that you need to make it to Heaven. Scholars will spend a great deal of time investigating all writings of the past, but the everyday person does not have the time to devote to religious studies. Scriptures speak of many names that are unfamiliar today. Ancient references to unknown names and places are confusing and can cause first-time readers of the Quran to lose interest in reading further. Humans use modern vernacular with recent geographical names. These days we use fewer rhetorical questions. The Holy Quran explains many concepts in the reverse of the way we think in modern times. The rhetorical question is not useful in media or print because it takes a moment to think of the answer. We live in a fast-paced world of instant communications; content bombards us from all sides, so we react instantly.

When ethnic or tribal conflict exists simultaneously with varied religious beliefs and politics, there is bound to be violent behavior. There is always a distaste for those who have different traditions and who speak other languages, so an effort is required if people are to reel in their emotions based on first impressions, for all societies have a wide range of behavior patterns.

A very peaceful sect of Christianity known as the Religious Society of Friends, or Quakers, believe that it is unnecessary to rely on clergy, liturgy, or creed. Quakers are aware that divine essence dwells within us in some measure, and that life is sacred and interconnected. They consider the revelation of God's truth as continuing and ongoing right up to the present day.

Members of the community come together in silence, emphasizing the element of togetherness without rituals or preachers. Each person is capable of direct unmediated connection with God.

Quakers welcome truth from whatever source it may come. When they assemble, it is in a spirit of waiting for inspiration from within. This open minded approach is refreshing, since many other denominations do the opposite by moralizing while preaching fire and brimstone to generate excitement instead of displaying a calm confident faith that never wanes. They do more than talk about being good people; the strength of faith leads to doing good within the community as the byproduct of love for our Creator. Actions speak louder than words. They believe that modelling God's presence in our lives is more important than espousing dogmatic or mystical beliefs.

The next chapter deals with a deadly threat to mankind – the Mark of the Beast. Those who accept the Mark cannot enter Heaven. There are many misconceptions on the topic, so it is important to understand the insidious nature of the Mark, and which institution represents the Beast. Members of that institution are people like Bill Gates, the Bush family, the Clintons, George Soros, the Rothschild family. The wealthy elite billionaires who work through secret societies such as the Illuminati, Freemasons, the City of London banking cartel, the Federal Reserve, Bilderberg Group, Council on Foreign Relation, and the United Nations are all controlled by the Jesuit Army of Loyola.

The Jesuits are the pope's private army of fanatics who engineer all the wars and terror attacks. Their long-term objective is to place the pope in charge of all nations, all material wealth, and all souls.

The Council of Trent is still in effect, so the plan is to introduce more genocide on non-Catholics. The Inquisition never stopped. The Jesuits take an oath to murder all enemies of the Vatican. The ongoing hoax of a pandemic shows how many civil service positions are occupied by members of this cult.

The plot is to destroy the souls of as many people as possible to keep in step with their secret Luciferian agenda.

Bill and Melinda Gates are Catholics who promote depopulation of the earth by injecting people with toxic vaccines that sterilize populations. The vaccine rollout is a genocide program that has already caused widespread mayhem in the Third World. The elites are afraid that the colored people will outnumber the white people, so they plan to kill of as many racially unacceptable people as possible before someone stops them. Bill Gates should have been put in prison a long time ago, but money talks and justice can be bought in most of the world. Now the Western Hemisphere must endure the brainwashing campaign over the fake virus pandemic that has hypnotized populations into a state of constant fear. The New World Order plan to divide and conquer the nations is coming to fruition.

Masks are an occult symbol of submission and have no purpose or effectiveness in stopping the transmission of disease. I was spurred to write down my feelings after seeing bearded long haired Orthodox priests addressing crowds on the streets of Toronto about some issues, yet the priests were wearing their polyethylene pandemic masks. There was no talk of faith in God, nor any condemnation of government lies; I was witnessing blatant capitulation to tyrannical policies that should have been ignored by those who pose as wise men. Those priests are obviously acting like hypocrites and fools in public.

Mysticism has crept into many religions, causing practical thinkers to avoid religious teachings altogether. The rituals appear distasteful when seen from outside by people who do not want to participate in such ceremonies. Traditional religion turns people away from belief in God since churches do not provide any answers for them, only more questions. The pastors, priests, and preachers speak as if they were authorities when they are just speaking as moralizers who claim to speak for God. The congregations develop into social clubs where some people believe in God and others simply believe in being good without acknowledging their Creator in true faith.

Priests treat their robes as if they were holy objects, kissing the garments as they dress. During the rituals, they bless some water and sprinkle it around to make other physical objects 'holy', as if they had some magic power. Orthodox priests attach divine characteristics to their beards, their hair, and numerous physical objects. This is idol worship, which is a mortal sin.

Leading others into pagan behavior or false doctrine is an unforgivable sin, for the wolves have spent their lives leading the sheep into destruction. Churches must dispose of their crucifixes, statues, altars, costumes, and icons if they are going to save anybody. We live in the present, so there is no need to dress up in garments from the dark days of paganism. Modern clothes will work just fine when people gather to honor God. The Catholic church in particular is a closed society - a system constructed on lies, deceit, idol worship, and superstition. The official Catholic 'church' consists only of the clergy; their literature does not consider the laity as part of the church, so the leaders will not hesitate to sacrifice their own followers if it promotes the greater good of the Vatican. Catholicism is a warmongering political force, not a religion.

Mankind tends to prefer multiple gods and goddesses that do not exist – man made myths such as the Immaculate Virgin Mary, Santa Claus, or the holy ghost with no name invented in the days of

Constantine. All those beings are supposed to be able to read your mind. The reliance on mythology has caused great confusion among those who are on the borderline of insanity due to other reasons. Uncertainty as to how many gods are involved has driven many people over the edge, since the ideas presented by trinity churches are impossible to reconcile. Trinity churches are luring souls into Hell.

The Mark of the Beast Decoded

People who know the Bible have seen the phrase, "These things are hidden in plain sight", so logic suggests that the Mark of the Beast is visible as something innocent looking today. Most scholars and students of Scripture are expecting some sort of fascist dictatorship coming along in the future, and a hypothetical world leader who would match the definition of the Antichrist. The common thinking is that all legal citizens would not be able to buy or sell without the mark. Current analysts think "the Mark" might be an RFID chip in the hand or a tattoo on the forehead bearing the number 666, the number of the Antichrist. The criteria of these two theories fall short of the elements necessary to qualify as the Mark of the Beast, however. The Mark of the Beast involves deception and worshipping the false god of the Antichrist. RFID chipping and forehead tattoos are not deceptive, everyone knows what they are for. Chips are obsolete with the advent of fingertip scans, and the rich would object to a tattoo with an inventory number. The Bible states that the Mark of the Beast would have something to do with the right hand, the forehead, the Beast and the number of his name. All these factors are in plain sight today. The Antichrist definition is the leader of the world's largest belief system who would try to make one world religion under his rule. This character is the Jesuit Pope Francis. The pope's Latin title is "VICARIOUS FILII DEI" (Vicar of the Son of God); the Roman numerals add up to "666". The Beast is the Vatican, and the name of its god is "Trinity", so I calculate the number of his name as three. The Mark of the Beast is the Sign of the Cross. You take an oath to the Three Unclean Spirits of Revelation: take your right hand, touch it onto the forehead, and chant, "In the name of the father, the son, and the holy ghost", thereby breaking the First Commandment to believe in One God and to have no other gods.

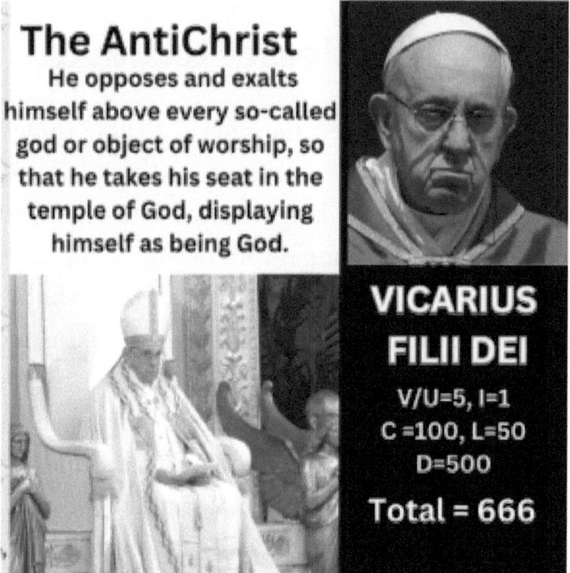

The dogma of the trinity is the entrenched doctrine of the Nicene Creed which must be discarded as man-made blasphemy, mere inventions of superstitious pagan cultures. Those who insist on dividing the indivisible God into three persons have taken the Mark of the Beast and are not qualified to enter Heaven. Trinity worship is Pantheism, consciously thinking of more gods, praying to those names or images, and the veneration of saints elected by men. The attempt to contact the dead is forbidden in Scripture, and it is mortal sin to make up other names that a person should look to as gods or goddesses. Mortal sins are praying to any entity other than God: praying to statues, praying over relics, using carved altars, chalices, incense, candles, robes to kiss as holy objects, necklaces, scapulars, rosary beads, icons, paintings, shrouds, and shrines are all idols. The Catholic doctrine does not include the Second Commandment against idol worship. The Bible condemns priests and promises that they shall be given the most severe of punishments. No man has the power to forgive your sins, and priests should not be called "Father" or "Holy Father" for that title is reserved for Yahweh, who is also called "Allah".

Decoding the Book of Revelation takes a lot of cross-referencing. Toronto police are looking for a motive in the April 2018 van massacre by a totally normal software programmer. The killer Alek Minnassian grew up in the North York area where he made the attack. He has no history of deep religious affiliations: he came from an Orthodox Christian family but was not actively religious. He had no political stance or affiliations that would inspire him to violence, just a note on his Facebook page that he was a fan of Elliot Rodger and the Incel movement (Involuntary celibate men who are rejected by women). Minnassian's mission in the rampage was revenge against women; this could be called a motive, but anger has root causes that lie in the human subconscious mind. Desire is the cause of suffering, so the problem is more than superficial rage. Sigmund Freud explained how the ego is compelled to act out the powerful repressed emotions of the id, but the commands of the id are balanced out by the superego's feelings of being weak, shameful, bad, and guilty. The ego performs a person's logic, so when there is a short-circuit in the psyche, the demands of the id will come to the surface striving for resolution. The breakdown in logic is caused by the Mark of the Beast causing a severe wrinkle in the person's ability to define God in a way that makes sense. The trinity deception is a virus eagerly spread by churches. It is the worst public health threat possible because the polytheist cannot achieve Paradise on Judgment Day.

The ancient pagan Babylonians recognized the doctrine of a trinity, or three persons in one god - it appears from a composite god with three heads forming part of their mythology, and the use of the equilateral triangle, also, as an emblem of such trinity in unity. The Hindu texts also present a holy trinity, The three gods, Brahma, Vishnu, and Shiva." Learn, O devotee, that there is no real distinction between us. What to you appears such is only the semblance. The single being appears under three forms by the acts of creation, preservation, and destruction, but he is one." Many other areas had their own divine trinities. In Greece they were Zeus, Poseidon and Adonis. The Phoenicians worshipped Ulomus, Ulosuros and Eliun. Rome worshipped Jupiter, Neptune and Pluto. In Germanic nations they were called Wodan, Thor and Fricco. The ancient Egyptians, whose influence on early religious thought was profound, usually arranged their gods or goddesses in trinities: there was the trinity of Osiris, Isis, and Horus, the trinity of Amen, Mut, and Khonsu, the trinity of Khnum, Satis, and Anukis, and so forth. The application of this old pagan conception of a Trinity to Christian theology was made possible by the recognition of a god with no name, the holy spirit as the required third 'person,' co-equal with the other 'persons'.

The idea of the 'holy spirit' being co-equal with God was not generally recognized until the second half of the Fourth Century A.D. In the year 381 the Council of Constantinople added to the earlier Nicene Creed a description of the Holy Spirit as 'the Lord, and giver of life, who proceedeth from the Father, who with the Father and Son together is worshipped and glorified.'

An organized thinker cannot resolve the doctrine of the trinity because it is nonsensical. God is all-powerful, omnipresent, invisible, and indivisible. Most of the 36,000 cults of Christianity insist that

Jesus is God, the holy spirit is God, and God the Father is God. The church elders voted on the issue in 389 A.D. and proclaimed the trinity as a universal truth. The church elders ignored the many third-party references to the Father by Jesus of Nazareth: Jesus stated in the Book of John that he was the son of God. It is doublethink to say that you can cause God pain and execute him on a cross - God would simply stop the pain; He would not stand for the disrespect and humiliation of crucifixion. Jesus never told people to pray to him; Jesus said to pray to the Father. When Jesus was baptized, a voice from Heaven called out "*This is my son who I love, in him I am well pleased.*" God has no superior; He prays to no one. Those who call Jesus God are incorrect. Only the Father can hear and answer prayers. God has all the power. People have nothing logical to believe in anymore. Many lose faith altogether: pop scientists like Stephen Hawking and Neal de Grasse Tyson tell everybody that there is no God, no Afterlife, no soul, and no such thing as fate. This is convenient, for it relieves people of all spiritual responsibility. Schools are telling students that life is accidental, meaningless, and temporary. Television has brainwashed the public to accepting replacements for the One true God of existence. Most educated people know better, so they turn into atheists or occultists as they give up on life. They feel no need to obey the Ten Commandments. This loss of ethics results in murder, theft, and disregard for the truth.

Many atheists are so confused they believe in Heaven and Hell at the same time they deny the Creator. People blame God for troubles out of anger over their own situations, as if cursing God would be an act of revenge. Those who have accepted that they are not worthy of Heaven no longer care what happens to anyone else. Retail religions have exploited the masses for money. The institutions of religion are so blatantly greedy they drive many people away from faith of any kind. The superstitious rituals that seem to be acts of piety are just shows put on so that people can punch the clock once a week, leave their donations, and return to the rat race without the slightest thought of showing gratitude to the God who created all things.

It is arrogant to believe you brought yourself into existence. The rich elite of power brokers are aware of their evil desires. They own the media outlets and use subliminal messaging to program the minds of viewers using phallic symbols such as the Saturn Five rockets shooting three men to the moon when two would have been enough. The triune deception is visible, yet invisible. There was an ulterior motive for the televised moon mission hoax: the visual image of three astronauts is a subliminal message that you should believe in three gods.

The erect missile rolling out to the launch pad is a moving phallic symbol that awakens the latent homosexuality of the subconscious mind. The Apollo team represents three-way authority, planting the image of a trinity in the sky. Other subliminal messaging is done by Hollywood - the Force, Darth Vader, and Luke Skywalker are trinity substitutes. Jesus is not God; the Messiah was the first of God's creations.

I am the oldest soul in existence after God, who always existed. Those familiar with my teachings from my life as Buddha will know the transmigration of the soul after death to a new body. Our many lives are like a row of candles, where each new candle is lit with the flame from the previous life. Many people will have talents from previous lifetimes that appear to be inborn rather than acquired, and this is due to reincarnation. The rebirth of the immortal soul is the basis of life. Your consciousness would get bored very quickly in eternity in you run out of data, the experiences that constitute the experience of living.

I am subject to the conditions of my birth as the reincarnated soul of Christ, so the things I do in life lead to predestined objectives. All things were created for me as it says in the Bible, for I am God's right hand man and the first born over all Creation. The human race was chosen as the species I would walk among in all my lifetimes, which implies that a continuity is necessary to achieve the proper development of the soul, rather than being placed into different civilizations around the universe. When we consider all the evil people in the world, it could be that the fallen angels were sent down to be human beings as well, engaging in the battle between right and wrong. It would be awful to think that other civilizations are as violent, greedy, and warlike as humans.

The nations have been fighting with each other for ages, so we should strive for a golden age of secure peace where all people can enjoy the time remaining until Judgment Day. The faithful have no fear of the inevitable end of mankind's physical existence, sincc it will mean entrance into the eternal bliss of Heaven and freedom from the drudgery of life. Most of the world's population are disbelievers who will scoff at any such notion of life after death, so they will find the truth after they are sent to be punished forever in Hell. The atheists are very bad gamblers, for their pride has caused them to bet everything on an empty pot. Those who have misled others into false religions to enrich themselves financially are just as foolish, thinking that God does not know their intentions.

Facts About Judgment Day

Institutional religions resist any changes to their set doctrines, so their adherents get more fanatical all the time in their elitist disgust for all other religions, but the truth is that God periodically sends me to introduce new covenants to mankind in various places and times. Like all messages that pass through the hands of scribes, translators, and church authorities, the words become partially corrupted with misunderstanding or bias over time. Eschatology is the term used for discussing matters concerning the End Times, and people should take heed of the latest statements I bring from God about the last things you must know.

The human ego is so bloated and arrogant that many people feel they will judge themselves as worthy or unworthy. These people will not allow any laws to guide them, nor do they accept the idea of Divine Retribution...they decide for themselves what is right and what is wrong. Most people do not have the slightest concern about the fate of their eternal souls in the Afterlife. Humans are irresponsible. Scriptures of many monotheistic traditions describe a Final Age, or End Times in the Judeo-Christian writings, The Last Day in the Quran...the Day of Resurrection and Ascension where the souls of the living and dead will arise to face Judgment. This is the event that will determine in which state your eternal soul will exist- reward or punishment. God has assigned me the task of being the first Judge since the Messiah has seen life from the bottom up while earning divinity. You might say God's perspective is from the top looking down, for His glory is supreme: by definition, everyone else is beneath Him and must earn favor by asking for enlightenment and seeking God's mercy.

God considers atheists and polytheists worse than murderers, for they fail to acknowledge their Creator; mankind has no faith; people are lovers of money, lovers of themselves, self-centered, and stubborn. Other well-intentioned people insist on following the dogma of churches, they believe that God came down and took human form as Jesus, as if God could feel pain. This is against what is said in the Bible; it is a common Heresy among sectarian interests. God the Father (Yahweh, Allah) and Yeshua the soldier, scribe, and messenger are two different souls. There is no trinity, no holy ghost - there is the Father and the Son. God can exist without me; I cannot exist without Him. Scriptures involving the fate of mankind in the End Times tells of hardship, famine, earthquakes, floods, and disobedience of the population to the point of a final War of Armageddon before the Last Day. Truly righteous people should have no fear of Judgment Day.

Evildoers think they can outsmart God somehow. God has given the task of Judgment to the son. I will be the First Judge and God can overrule me if any complexities exist that might change the verdict. "I am the Way, the Truth, and the Life. No one comes to the Father but through me." If you have prayed to Mary or called upon other false gods, there will be no reward waiting in the Afterlife. Only the purest of souls can be allowed into Heaven on Judgment Day.

Praying is to be done in silence in the privacy of your room, not on street corners or in churches. The social club functions of churches make it hard to disengage from a trinity religion. A wise person will not succumb to peer pressure, however. It is better to satisfy God than to satisfy your fellow man. Purity of Intention runs the show, not some professional preacher or priest. They know only what they have been programmed to know. Priests are complacent because they live away from reality. Their houses are free, their bills are paid, they live in comfort and luxury at the expense of their victims. They will never change, for they have been told the ends justifies the means.

The priests teach a nonsensical doctrine that a person lives only once and is Judged immediately after death. This is a dangerous lie believed by most of the Christian sects without question. There are specific Bible verses describing Judgment Day: Matthew 25: 41 "Then shall he say also unto them on the left hand, depart from me, ye cursed, into everlasting fire, prepared for the devil and his angels: And these shall go away into everlasting punishment: but the righteous into life eternal."

"And whosoever was not found written in the book of life was cast into the lake of fire."

The many commercial preachers and others who claim God speaks through them are frauds who will pay the ultimate price for misleading others. I alone will speak the words that God has ordered me to speak. All prophets are forbidden by the Bible in multiple verses:

Jeremiah 23:11 "I did not send the prophets, yet they ran; I did not speak to them, yet they prophesied."

Deuteronomy 13:3 "You shall not listen to the words of that prophet or that dreamer of dreams. For the Lord your God is testing you, to know whether you love the Lord your God with all your heart and with all your soul."

Islamic and Christian leaders of today make great efforts to deny the facts of reincarnation, calling the notion of a re-born soul "a false science" when the ones making these statements are not scientists. If any group continues to spread hatred against the truth of reincarnation, they will be rebuked as a false religion not authorized by God or the Messiah in any way. Islamic leaders must consult with me and make the necessary changes if they want to be called the true religion. Those who disagree with truth are not worthy of Paradise, and they shall never be with God in eternity. My words are the only truth. I must authorize any religion, or else their members will not pass Judgment. Those people who continue to call on the trinity or Mary will be sent to Hell.

God knows your thoughts, so there will be no chance to perform the pagan rites of the past in secret. The priests and clergy of all religions must cast away their silk robes and remove all idols: the objects of idolatry are chalices, crucifixes, relics, statues of deities, communion wafers, candles, and incense. No one may attach spiritual significance to long hair and beards. Priests should not hide their faces behind a façade of holy hair. Remember that idol worship is mortal sin. Those who love the things of this world are the enemies of God. Mankind always tends to return to idols and multiple god figures.

Skeptics and pundits may ask "where is God?" to taunt believers in debates, thinking the question will be impossible to answer. Educators have dropped religious teaching for the most part, so young people are not given any definitions for the Supreme Being.

It is important to understand that God is omnipresent, although we point up to the sky as a first impulse. This sense of a higher position is not physical; the invisible dimensions are higher planes of existence. God's presence is inside us and outside us, but His consciousness is concentrated much like the eye of a hurricane and located at what we call the Godhead or Heaven, the central dimension. The material universe is only one of ten dimensions of existence.

After Judgment Day, Heaven and Hell for mankind will exist in the Fifth Dimension. Existence for the conscious souls of all humas will be in one of two places – Paradise or Hades. On the Last Day, each person's history and thoughts will be known and judged accordingly. Atheists and polytheists will not be allowed into Heaven, for they have refused to acknowledge the One true God. People who have victimized others will relive the agony that they have inflicted, and their punishment will never end.

After the Judgment, Jesus Christ will lock the gates of Heaven and Hell permanently. There will be no escape, no parole, and no release from the devil's furnace of darkness and fear. The souls who fail to attain Heaven will exist in torment for eternity, regretting that they were ever created. All people on earth have had the opportunity to know the laws of God, but they chose to disobey the Commandments and they ignore the teachings of Christ.

It seems like an easy choice to make - people can strive to be good, thus attaining the reward of Heaven, or they can throw it all away by lying, stealing, or killing. People routinely condemn themselves by displaying disbelief and scorn for God. When a person spreads atheism to another, he has destroyed the future of all who believe him. Egotistical people will laugh at the idea of God out of ignorance.

A fate worse than death is falling out of God's favor. The souls who get sent to Hell will never hear God's voice; they will experience an eternity of fear, humiliation, pain, and loneliness that is too horrible to describe in words. Once that consciousness becomes aware that the punishment will never end, then a panic will seize the sinner and engulf that person in total despair.

The sixty-six books that are included in today's Bible version were compiled long after the time of Christ through numerous authors, and the officials of the Roman Catholic church held editorial control for centuries. Some Biblical passages express universal truths that are morally valuable and wise no matter who wrote them down. Treating others as would treat yourself is sensible, yet mankind generally maintains a cruel streak of hostility towards other people and indifference to the suffering of others. Most people would agree that money is the root of all evil, then we see a world full of very wealthy televangelists, preachers, and self-proclaimed prophets who rake in tax free money to pay for their private jets, luxury cars, Rolex watches, and mansions. The commercialization of religion is rampant, and well-fed by the donations of insecure followers who have been taken in by the fraud artists who speak from the stages through forked tongues.

Although the Bible used today has many contradictions after sixteen centuries of Vatican editing, it has astonishing prophecies that tell about the coming of the dictatorial forces that we see running most of the world in these times. The situation is that a fascist cabal of unelected rich people are trying to gain power above sovereign nations by declaring fictitious world emergencies in matters of health and the environment so that they can remove all personal rights. Those who are in police and military positions would be wise to resist before they are deemed redundant and replaced by United Nations personnel.

Our existence as conscious beings is highly unlikely and against the mathematical odds, but nevertheless we exist and look out at a world that surrounds us. Most people take this phenomenon for granted without giving it a second thought, but those who are attentive will question how it came to be. All people are forced to navigate through society in some way, and the journey is not always smooth sailing. The important thing to learn during the trip is the nature of our destination, which will be a spiritual state of existence rather than a physical place.

It is human nature to desire that things should always go the way we wish, so we become dissatisfied when we get attached to temporary things such as possessions, people, power, and life itself. Unfulfilled desires cause suffering because the things of this world will inevitably pass away at some point, so the way to stop suffering is to detach oneself from desire by learning self-awareness, humility, generosity, and self-control. Once the ego is defeated, then the unhappiness of attachment will fade, and you will feel more satisfaction when going through life.

It takes effort to build up the spiritual discipline search for knowledge and to study with an open mind, for those who cling to set beliefs for the sake of conformity will lose the race for personal salvation due to their irresponsible laziness. We live in a busy world full of distractions, so very few people find time to think about what is going to happen to them in eternity. Society has become a materialistic rat race where all members must compete against each other, propelled by the fear of being stuck without money. People must think beyond the activities of this world and learn to separate the temporary things from that which is permanent.

Listen, I tell you a mystery: We will not all sleep, but we will all be changed - in a flash, in the twinkling of an eye, at the last trumpet. For the trumpet will sound, the dead will be raised imperishable, and we will be changed. For the perishable must clothe itself with the imperishable, and the mortal with immortality. When the perishable has been clothed with the imperishable, and the mortal with immortality, then the saying that is written will come true: “Death has been swallowed up in victory.”

Quantum Entanglement

The Local Theory of classical physics implies that objects are only affected by their immediate surroundings, yet many people have the instinctive feeling that life is affected by intangible factors such as fate. Particles are influenced in their spin, momentum, and position by forces and particles that exist at a distance, and so consciousness is also connected to other phenomena throughout space and time. The term Quantum Entanglement can be applied to Psychology of the Masses. Most people are not math enthusiasts, so you might wonder what these theories have to do with the price of rice in China, and what does this mean to the average person. We are in a stream of data, so if the source is not providing truth, the receiver is liable to succumb to Dissociative Identity Disorder. Quantum Entanglement means simultaneous existence at different points and includes conscious experience plus the remembrance of many experiences at once. The impact on a human soul is important, since in the Afterlife your consciousness will be experiencing either pain or pleasure similar to that which we experienced in the physical world but magnified to the extreme. We can feel a certain amount of pain and pleasure during dreams without the benefit of a physical body, so it is logical to conclude that there will be feelings beyond this dimension.

Clues to the detachment from actual reality manifest themselves in the way each person perceives reality - what we take in and accept as truth is set up by someone else in another land and another time, an idea that pollutes logical thought processes remotely. Here is an example: ten subjects were shown pictures of several well-known paintings, then I took their feedback individually without asking any questions. The four paintings were "The Last Supper", "Mona Lisa", and "Portrait of Jesus Christ" by Leonardo da Vinci. The fourth masterpiece was "The Scream", done in 1893 by Edvard Munch. One person commented who was a male, and a devout Catholic. He failed to recognize the Mona Lisa and said it was the Virgin Mary. A female respondent mistook the Jesus painting for the Virgin Mary. Those two did not like the Munch painting (which sold at auction for $120 million). They had never seen it before. They called it a child's bad drawing, a dumb cartoon, the work of Satan. They both called Leonardo a heretic. Religious zealots have the habit of denouncing and judging other people. The other respondents were not Catholics, and they did not see Mary in any of the paintings. They identified all four paintings correctly.

The educational programming of separate schools imprints a false reality into the minds of unsuspecting victims. Many religions use the mythology lever to control their members, instilling fear of rejection from the group so that few will question the doctrine being presented to them. Fundamentalist religions tell their members that science is the work of the devil, the earth is flat, and there was no Big Bang. Falsehoods are easier to sell than the truth. People hear what they want to hear, and humans routinely digest gossip as truth. Entire populations are intimately entangled to people and places from the past, from the traditions they initiated out of ignorance. The intolerance of powerful despots of the past lives on through the traditions built up around them, into ripples of hate and ignorance that reverberate to this day.

Misery and suffering are the end products of reactionary schemers who act out of greed and speak out of ignorance. There is an entire prophet industry today where there should be none. Snake oil salesmen turn religious quickly when they count the profits of pretending to be a prophet. The important thing for both atheists and religious zealots to understand is that life is not all that it appears. Metaphysical notions are not witchcraft or deception. The churches do not represent the policies of God; they only give interpretations of what they believe to be true, and they say whatever will work to please the congregation.

The populations of Western cultures have been conditioned to be entertained by action and violence. The result is that people do not want to hear about reincarnation or an eternal Afterlife: they are trained to become sick and tired of life. Most people expect nothing after dying but oblivion. Too many people assume they have a right to Heaven, and everybody with a big smile goes to Heaven. Humans have a habit of insulting those who disagree with them in any way. Physics says every action has an equal and opposite reaction. That goes for the consciousness as well. Evil intentions make ripples that can come back on you. What goes around comes around. A conscious soul can be in many places at once under the right circumstances. The great manifestation of Quantum Entanglement for me and the entire human race will occur on Judgment Day.

Scripture describes the Judgment as being The Last Day, so I will be simultaneously aware of all the thoughts and actions of everyone, facing all souls who have lived in human history. In the Afterlife, those who achieve Heaven will be able to re-experience happy moments from several former lifetimes simultaneously. The experience of simultaneous memory recall, re-experiencing many happy moments with their accompanying joyful stimulation is a sort of Quantum Entanglement type of reward - the gift of happiness and freedom from the material world. Paradise will be a limitless playground of constant excitement and joy. People routinely laugh this off as not important. Most humans today are atheistic; they take no heed of the many warnings of Scripture about eternal punishment versus the chance at near-limitless rewards that will never end.

The religious movements have entangled mankind with a built up a mythical version of Jesus Christ, portraying me as a hairy, bearded, victim screaming on a cross. Hollywood and Broadway have capitalized on the myth of Jesus Christ as a superstar or a homeless vagrant, and everything in between - anything except a dignified Zen monk who is quite familiar with pain and does not lift his voice or make it heard in the street. I am a man of constant sorrow, despised and rejected by mankind. Anyone who does not fit the myth is deemed to be a liar immediately by jealous people.

There are many people in Christian sects who insist that Jesus is God, and there is no way they will ever change their minds. They are locked into the lie. Since Christ is the soul of light, and light is finite, then Christ is not infinite. God is infinite. The New World Order propaganda machine of news media, television, and Hollywood movies has trained people to be materialistic, murderous, and confrontational. They use the tactic of 'divide and conquer' to set religions against each other by infiltrating the hierarchy of other religious groups. The Catholic church spreads gossip about Prophet Muhammad, not knowing that Muhammad was a later lifetime of Jesus Christ and Buddha. When you insult Muhammad, you are insulting Jesus. The desire of the Catholic church is total control, the New World Order that Pope Francis promotes so enthusiastically. Tens of thousands of pedophile cases are active against Catholic priests, so they call other people pedophiles to put up a diversion to hide their own crimes.

The pagan idol worship of the Catholic church has been forced on Third World countries. The church wants control over body, soul, and all nations. Many opposing religions will not believe that I experienced time travel and Quantum Entanglement, being in three bodies in three different time zones. I can do nothing by myself however...the experience was engineered by God as an early reward. The experience lasted about ten minutes; in that short amount of time, I got a sense of increased wisdom.

To offer a true religious doctrine, you must take ideas from Buddhism, Islam, and Christianity that can be considered universal truths. All religions have been corrupted by the traditions of man. The message of the cross sounds foolish to those who are on the road to destruction. Those who wish to be saved must recognize that God has total power over life and death. That is the message of the crucifixion and resurrection. God deliberately chose things the world considers foolish in order to shame those who think they are wise. And He chose those who are powerless to shame those who are powerful. God's way seems foolish to some because they want a sign from heaven to prove it is true. And it seems foolish to others because they believe only what agrees with their own wisdom. As the Scriptures say, "I will destroy human wisdom and discard their most brilliant ideas. Where are the wise debaters? where are the disputers of this world?"

God has turned the wisdom of this world into foolishness. Despite being given the wisdom of God the world took wisdom, but they did not acknowledge the source, which is God. It pleased God to save those that believe by the foolishness of preaching. The wise men who are after matters of the flesh are not called, not many mighty, not many noble, are called. No person of flesh should glory in his own presence. God has chosen the foolish things of the world to confound the wise; and God has chosen the weak things of the world to confound the things which are mighty, the base things of the material world, and things which are despised; soon God will turn the material things into nothing.

The Creator God gives us a chance to acquire wisdom, righteousness, and redemption. He has arranged for the ethereal qualities, that which is pure consciousness, to continue throughout eternity. Both the churches of tradition and the televangelist-style preachers have a vested interest in perpetuating ignorance and continuing the falsehoods they spew to the gullible. They have mansions to upkeep, and they purchase drugs such as cocaine to keep them talking. The preachers live in luxury while poor people starve and have no medical care. Money is the god of the priests and preachers. You must gather spiritual treasures for yourselves rather than material treasures. The things of this world are temporary, but the Afterlife will last forever. This is difficult to visualize since there will be no end to existence. There is no point to holding money without helping others who are in need. It is better to give than to receive, for your generosity and good will towards others will make you stronger.

Most doctors, lawyers, and politicians I have spoken with have stated that they do not believe in God. Their training is specialized, so they have not studied subjects such as Logic, Philosophy, Theology, or Physics. If they were curious about the destination of the eternal soul, there is plenty of literature on the existence of God. Professionals like to think of themselves as self-contained experts who do not need to rely on a Supreme Being to take care of their future experiences. Those who are atheists do not care about Heaven and Hell at all; they do not ponder such issues.

Atheists are so zealous about telling people there is no God, they fit the definition of a religious cult, since belief systems are religions with followers. The eager beavers who line up for an untested vaccine to address a non-fatal inflammatory condition requires faith in corporations and governments. If you trust in large pharmaceutical companies and warmongering world leaders, then you do not trust in God. Humans have an immune system to take care of illness. Those who have faith in God will not wear a mask or take an experimental injection of chemicals and tissue from aborted babies. The vaccine cult is suicidal in nature.

Much like the Heaven's Gate cult in California led its followers to kill themselves because they thought an alien mothership was going to "save" them while hiding behind Haley's Comet, today's vaccine cultists believe they will be saved by the very pharmaceutical monsters that are hell bent on maximizing vaccine profits no matter how many people die in the process.

Over 4,000 deaths have occurred to people who have received the needle, yet the mainstream media writes off all deaths as coincidences, and they ignore the thousands of people who have been crippled and injured from the Pfizer, AstraZeneca, Moderna, and Johnson & Johnson versions of the gene-altering vaccine products.

The vaccine zealotry has gone so far beyond any realm of reason or "facts" that it can only be accurately called a cult. And it is a suicide cult, since so many are setting themselves up for mass death from the medium-term side effects that end in horrendous suffering, known as Antibody Dependent Enhancement or ADE.

This is also known as a "hyperinflammatory reaction" upon exposure to coronavirus strains in the wild, and it presents a very real risk of mass fatalities among those who are dumb enough to have lined up to take the experimental mRNA vaccine that has been shoved down everyone's throats through coercion, threat of job loss, and the deception of giving out gifts such as grams of marijuana or meals to volunteers who agree to take the inoculation.

The vaccine brainwashing campaign managers make every effort to censor, ridicule, and discredit anyone who dares raise even a single question about vaccine testing, efficacy, or testing. If anyone states that more than twenty European countries have paused the deadly AstraZeneca vaccine linked to blood clot deaths will get you banned as an "anti-vaxxer." No person is allowed to cite any safety concerns, even those expressed by other governments.

Dr. Derek Knauss works as a clinical lab researcher in California. He and his team of virologists examined 1,500 samples collected that were supposedly COVID-19 cases. They used a powerful scanning electron microscope but found no such virus in any of the samples. All they found was Influenza A and Influenza B, but no novel virus. The team concluded that COVID-19 is imaginary. No one has ever isolated such a virus, and the CDC documents show that they do not have a sample.

What we are dealing with is just another seasonal flu strain like every year. COVID-19 does not exist; the ad slogan virus is fictitious, made up by the globalist New World Order agents to bring in worldwide tyranny and a totalitarian police state. This conspiracy to overthrow all free nations includes plans to exterminate the population through repeated lethal inoculations disguised as vaccines.

With the lure of huge immediate profits and the future capability to manipulate human health, the promotion of vaccines has itself evolved into an institutionalized money pipeline. This marketing plan is designed to create a need for long-term care for its victims. With cold-blooded calculation, the corporatists supply health insurance, and medical services to profit from the suffering of the people.

Israel is a testing ground for mandatory vaccinations and vaccine passports. Those who do not accept the inoculation are not allowed to travel or to shop in stores. The intention of governments is to cut off all services from rebellious free thinkers and to lock them away from society totally. This is a medical Apartheid system, where all human rights will be a thing of the past. The state will eliminate unions and replace unionized workers with Third World immigrants so that the citizens will not have the means to resist.

This grave state of affairs has been put into place step by step, using the Swine Flu scare of the previous decade as a blueprint for future deception. The constant stream of lies from radio and television keeps the element of fear alive, so that people will demand a cure. They are lining up like lambs to the slaughter although over 4,000 deaths have been reported from the inoculations so far. The politicians pretend there is a vaccine shortage so that panic-stricken people will try to get in line faster.

Brainwashing is considered a war crime according to the Geneva Conventions and the Nuremberg Codes. It is a crime against humanity to administer inoculations to citizens without their informed consent. It is routine to give vaccines to newborn children, and many senior citizens in nursing homes are not competent to make such decisions, but they get double shots without the consent of the patient or the next of kin. Military personnel are given injections whether they want them or not.

The pharmaceutical companies offer their vaccines to governments because the orders are placed and paid for in advance, so profits are guaranteed. The laws get changed to release the pharma companies from all liability for causing crippling injury or death to those who receive the injections. Pharmaceuticals are a multi-billion dollar industry, so greed overshadows ethics or any empathy for the victims of their toxic chemicals. Bureaucrats are also investors, so they cash in on the rules they impose on a gullible public. The investors will make fast profits before the death and injury count caused by their products becomes intolerable; that would bring about a financial backlash.

Politicians make the bylaws and decrees that all must follow, and they know what is going to happen ahead of time. They buy shares, whether directly or indirectly, in the exact companies that supply the mandated products, such as plastic shielding in stores, signage, stickers, tester kits, masks, syringes, and printed products. The rulers pay the police, so they are never investigated for Conflict of Interest violations. The leaders have extended the lockdown over and over in most countries; there is nothing to stop them from suspending elections indefinitely due to the fake third wave outbreak. They have total power under the emergency; tyrants are not likely to give back such power voluntarily – the current leaders of the lockdowns must be forced out of office, arrested, and tried under military tribunals where they can answer for their crimes.

There must have been an overwhelming series of events leading up to the cooperation of those in high office who preach crude pseudoscience to justify their actions. Whether through bribery, drug abuse, or blackmail, the civil servants have lost all feelings of pity and human empathy. People will go hungry and homeless because of the endless restrictions on earning a living. The far-reaching medical edicts empower authoritarian governments, not human health. The World Health Organization operates like a global oligarchy, forcing all member states to carry out their orders.

Tampering with the immune system for every virus, especially after it has mutated is a waste of time and money except to the corporations that sell millions of doses to the governments in lush contracts so the state can brag that they have taken measures to contain a pandemic when in fact they are simply creating another national scare to move tons of worthless concoctions.

False Prophets and Dangerous Cults

The Bible predicts that many Jesus imposters and numerous false prophets would appear in the End Times. Many televangelists have taken to announcing that God speaks to them; others go all the way and claim they are the reincarnation of Christ himself. In Australia, there is a cult run by one A.J. Miller, who tells his follower he is Jesus. In Russia, an ex-traffic policeman runs a large Jesus cult in Siberia, where he is Jesus in flowing robes, long hair, and beard. The Philippines is plagued by the cult operated by Pastor Apollo Quiboloy, who also tells his followers he is Jesus Christ incarnate. In the USA Jimmy Swaggart claims God speaks to him; the late Kim Clement held televangelist meetings where he spoke as if he were God. NAR cult spokesman Mark Taylor, televangelists Kenneth Copeland, Robert Tilton, Joyce Meyer, Joel Osteen, and Benny Hinn all claim to speak on behalf of God. Prosperity Gospel preachers ask for money, telling their hypnotized fans that their "seed" will grow into riches later, courtesy of God.

The Mega-church pastors live in mansions, they buy private jets, and they stash away the millions of dollars that they steal from gullible people who believe them. All the people who claim to be prophets are false. People who claim to have had encounters with Jesus are liars.

Atheists are so zealous about telling others there is no God, they have become like the religious people they criticize. Some celebrities make a living scorning the thought of any Creator of existence - people like George Carlin, Bill Maher, Penn Gillette, Richard Dawkins, Rickey Gervais, and Peter Hitchens all made the big gamble by laughing at God and doing it on television to mislead others.

God has a "special punishment' in mind for atheists who mock God and for polytheists who turn to idols and humans for salvation. They have turned up their cards and wagered everything on an empty pot. A smart gambler will see that there is nothing to win by being an atheist. The only chance to attain a reward is to strive for Heaven.

It appears that people feel a need for another human to guide them. The Church of Jesus Christ of Latter Day Saints, known as the Mormons, follow the revelations and a totally man made gospel called the Book of Mormon. Mormons do not believe in the trinity concept of one God being three persons, however they do not believe in One God and one mediator; the Mormon church and its derivative sects believe that there are three separate gods called the Father, the Son, and the Holy Ghost, which is anti-Biblical, because God is spirit and He cannot be divided.

Joseph Smith, the founder of Mormonism, was born on Dec. 23, 1805 in Vermont, USA. In 1820, when he was fifteen years old, he claimed that he had received a vision that went like this: " I asked the personages who stood above me in the light, which of all the sects was right- and which I should join. I was answered that I must join none of them, for they were all wrong; and the personage who addressed me said that all their creeds were an abomination in his sight ."

He reported a second vision on Sept. 21, 1823 where "a personage appeared at my bedside" to inform him of the location of a box containing golden plates buried at Manchester, New York. In 1827 Smith claimed that he was led by an angel to the golden plates upon which the Book of Mormon is alleged to have been written. He claimed to have finished translating the plates on March 26, 1830. Smith said the plates were written in "reformed Egyptian hieroglyphics", and by using Urim and Thummim magic spectacles, translated them into English in 1830 as the Book of Mormon. No Egyptologist has ever unearthed a "reformed Egyptian" language. It has Roman letters inverted or placed sideways, mixed with Greek and Hebrew letters.

The Book of Mormon claims to be the history of 2 ancient civilizations on the American continent. The first group left the Tower of Babel about 2250 BC, crossing to Europe and sailing to Central America. They were totally destroyed because of 'corruption'. The second group allegedly left Jerusalem around 600 BC, before the Babylonian destruction and captivity. They sailed to Peru. These supposedly were righteous Jews, led by Nephi. The Mormon record claimed that Christ appeared to the Nephites, to preach the gospel to them, to institute baptism, the communion service, the priesthood and other mystical ceremonies. They supposedly later on split into two warring camps, the Nephites and the Lamanites (American Indians). The Lamanites allegedly wiped out the Nephites completely in a battle in Palmyra, New York, in 428 AD. The Lamanites supposedly were cursed with dark skin for their evil deeds. The preposterous stories in the Book of Mormon are racist fantasies and historical nonsense, but people will believe anything they are told.

On April 6, 1830, at Fayette, New York, the Mormon church was organized, with only six members. They then moved to Kirtland, Ohio, which is near Cleveland, and published the book "Doctrines and Covenants". Smith was imprisoned in Far West, Missouri, for fighting. After escaping, he and his people fled to Nauvoo, Illinois, where he organized a small army. When a local paper, the "Nauvoo Expositor", published anti-Mormon material, Smith ordered the press destroyed and the paper burned. This act of destruction led to Smith's arrest and imprisonment. He was sent to a jail in Carthage, Illinois, with his brother Hyrum. On June 27, 1844, a mob of 200 people stormed the jail, shooting and killing Joseph and Hyrum Smith. The Mormons consider him a martyr.

Brigham Young then took over the leadership. He led the group westward to found Salt Lake City, Utah, on July 24, 1847, which became their headquarters. When Young died in 1877, they had 150, 000 members. These days the Church of Jesus Christ of Latter Day Saints has six million people worldwide who follow the heretical false gospel called The Book of Mormon. They claim that Jesus Christ restored his church through Joseph Smith, who was nothing more than a liar and a wolf in sheep's clothing and one of many false prophets who have committed the unforgivable sin of leading people astray.

God is not Jesus, and Jesus will not be a divinity until after Judgment Day when Christ will be the King of Heaven, courtesy of his Father, who is Almighty God. The cults who insist that God came once to earth in human form as Jesus are following a tradition made by the Catholic church, who ruled all nation for centuries. The founder of the Catholic church was Emperor Constantine, not Emmanuel of Nazareth. They perpetuate the lie that Jesus founded Catholicism so that they can claim authority over all souls.

The largest cult based on lies is the Catholic church, with over one billion members who believe that the pope turns into the spokesman for God upon election. The priests perform sacraments for money, insisting they have magic powers to forgive sin and turn a wafer into Jesus' literal flesh and blood.

The doctrine of the Eucharist is not symbolic - the church member must believe the wafer has become the literal body and blood of Christ. Zealous Catholics convince themselves they hear a heartbeat going on inside the blessed wafer. Upon examination with a microscope, there is no change whatsoever in the wafer the offer as a holy sacrament. The church uses the communion wafer to extort obedience from loyal Catholic parishioners: if they disobey the church, they can be excommunicated and denied communion, a procedure known as "excommunication". The Dark Ages rituals, fish hats, and silk costumes persist in the idolatrous pantheism displayed by the Cult of Mary. The elitist and violent actions of cults are universal. The leaders censor all outside information that disagrees with established dogmatic practices. The Catholic cult remains in existence because they indoctrinate children. The Catholic church has been the cause of most European wars and world wars. Catholicism has a long history of elitist political rule, oppression of all who dared speak against its pagan doctrines of blood sacrifice, superstition, multiple gods, and material greed. The

Inquisitions of history were sadistic incidences of ongoing terror against the poor. Crusades were the result of Catholic aggression and the Vatican's stifling autocracy. The popes have an army, the Jesuit Army of Loyola. They work under an Oath of Induction. They promise to act as spies and infiltrators in all social positions in order to enforce the will of the pope. The Jesuit Oath has them promising to hang, waste, boil, flay, and mutilate those in other religions, to "rip the babies from the mothers' wombs and to crush their infants' heads against the wall in order to exterminate forever their inexorable race." Jesuits have infiltrated all countries in all levels of society.

Their operations are not the activities of God-fearing men, they will commit genocide on order. In 9th Circle rituals, bishops meet to sacrifice babies and drink their blood. The cardinals and bishops must take part in these crimes to be promoted. The Synod of Bishops pray and sing to Lucifer; they are Satanists in secret. For centuries, the priests have hidden under the cloak of respectability, yet inside they are ravenous wolves stalking their targets and molesting children at every opportunity. Sam Giancana said in an interview that Vatican City is Mafia headquarters. The diplomatic immunity of the Nation-State of Vatican City allows them to smuggle suitcases of cocaine, launder money for the mob, and run child sex slavery rings, where they abduct children and sell them to rich elites around the world. The god of the Catholic church is the Almighty Dollar.

The priests are a homosexual cult of child molesters pretending to be holy men. They are perverts and parasites that should be banned from society. The doctrines of Catholicism are totally contrary to the Bible and a guarantee of failure on Judgment Day. The Catholic church has erased God's Second Commandment and they ignore the First Commandment to have One God. Other religions such as Mormonism, have adopted an entirely different gospel contrived by men. Humanity must learn to understand the truth and put aside their zeal for bizarre religions if anyone is to see Heaven. There are many self-appointed 'prophets' today with large followings, so it is obvious that people are looking for new Scripture when they haven't analyzed or understood what is already in the Bible. It is forbidden to add to God's words or take away from words from Scripture unless you have authority.

God has voiced His opinion to me on the current condition of humans: The Creator is angry and disappointed with mankind's failure to develop into peaceful and cultured people. God describes humans as "self-centered", "egotistical", "ignorant", "irresponsible", and "too stupid to memorize the Ten Commandments". Humans would rather jump into the mouth of a dragon than to admit they are wrong about anything.

The concepts in this book will help some people to shake off the yoke of false doctrines put forth by greedy religions. Faith is between the individual and God. The constant drive to look to other humans for salvation is self-defeating. Priests, pastors, and preachers have taken up a profession of pretending to have magic powers and a mysterious special connection with Heaven which they do not possess. The clergy lives as an elite higher class; they live lives free of debts and they do not have to pay taxes. Their trade is selling pipe dreams to people desperate for knowledge. Most churches are intolerant elite groups who expend great effort to criticize other religions. These competing religions can only lead to destruction, for religion was a basic cause of most wars throughout history. Organized religions are a major threat to the individual human being who wants to know the truth; churches must insist they are right in order to survive financially.

The Catholic church has mastered the art of brainwashing and censoring out all outside information that contradicts their elaborate doctrine. The Vatican's Jesuit operatives hold ownership of the World Health Organization, most major publishers, radio, television, and motion pictures. The announcers and hosts churn out propaganda to rationalize all the many falsehoods and discrepancies in their literature.

The Catholic system allows men to vote any person's name into divinity, as a saint after death unilaterally. Mary was made divine in the fifth century by the votes of men, not by the Scriptures. The fantasy world of revised history blemishes every area of Catholic thought. The church uses brainwashing techniques such as endless repetition of lies, hypnotic music, blocking of outside information, and threats of excommunication to force their members to toe the official line of doctrine.

Several large religions advertise that there is no salvation outside of their particular religion, so they lie. Catholics are told to believe they cannot be saved without the weekly communion wafer. In this manner, members are programmed to believe they must take communion; the ritual effectively holds the parishioner hostage to the mystical cookie. The pagan rituals and dark ages costumes of Catholicism provide an escape from reality for entranced churchgoers. Karl Marx said, "Religion is the opiate of the masses." Zealots will insist they are right in the face of mountains of contradictory evidence. The pride of man is its downfall. The human race has been a failure in the eyes of God, yet no one seems worried about engaging the wrath of God. People have become convinced that they can propel themselves into Heaven without consent from Jesus Christ or God. That is the height of egotism.

Those who do not try to learn how to attain Heaven are not likely to find God. When danger appears, people today do not turn to God, they turn to their phones to make a final entry on social media to say, "good-bye cruel world". Make a mental note not to worry about your online reputation when the plane is about to crash. That is the time to say your prayers, ask forgiveness, and ask to be allowed to live or to get into Heaven if death comes upon you. Social media cannot save your soul from damnation. Too many people assume they can place themselves in Heaven unilaterally and this is a mistake made by the gigantic human ego.

Existence is governed by the Creator and all beings must follow God's rules or else they will suffer the consequences of disrespecting God and challenging His infinite power. The intentions of your mind can be sins of wishing harm to others. The fact that one is not able to act or is fearful of prison scarcely matters. The spirit of most religions is for all people to go about their daily business without interfering with others. People still engage in stealing without considering the sinful nature of depriving others of their money. The first reaction of a righteous person is to show empathy...put yourself in the other person's position. You would not enjoy being insulted, robbed, or ridiculed, so it is not constructive to have the habit of oppressing those you find distasteful. Taunting the old and the poor will not get you to Heaven.

The churches have no authority from God, yet they chastise others and tell them to repent. The priests and the preachers cash in on the suffering of the messiah's crucifixion by collecting money from people who feel guilty about lynching Jesus. This sympathy card is used extensively by the carnies who go into the religion business as an alternative to making an honest living at a real trade. There are no standards for truth in a political climate of "religious freedom". You have no freedom to make your own religion in the eyes of God. The government will allow a preacher to lead the flocks into damnation with their lies. It is worse than murder when a preacher or priest tells you there is a trinity of three god persons. People begin to address these imaginary beings in the third person, as in "The holy ghost, he fills my heart with joy" or such other polytheist thinking. God exists as spirit, so the holy spirit of God is not another person. God has said in the Bible that mankind always tends to go toward idol worship. Looking to an image some artist has portrayed as Jesus of Nazareth, then treating that image as God is idol worship.

People must treat God as the unseen infinite Creator, not a naked corpse hanging on a cross. The human obsession for pain and death is very pronounced among religious zealots and atheists alike. The history of mankind is tarnished by numerous wars, acts of capture and slavery of other humans, cannibalism, sadism, oppression, and robbery. Liars, thieves, and murderers are treated as heroes in the movies and on television shows. People are generally intolerant around the world, so this must change. Human society always reverts to idol worship, as they revert to cannibalism when they are starving.

There are numerous crimes being committed by governments and elite cabals, who are in positions to cover up the truth about child sex slavery, embezzlement, murder, fraud, war, and other crimes against humanity that have gone unpunished by earthly courts.

"Then the Lord saw that the wickedness of man was great on the earth, and that every intent of the thoughts of his heart was only evil continually."

The End of Mankind's Physical Existence

The elitist cabal which controls the media and governments base their secular ideology on the assumption that the human race will continue for centuries, and they spend money today on programs that take funds away from citizens in the present to work on a hypothetical future. They do not talk about radiation leakage into the ocean from Fukushima or the constant threat of World War Three. Forces within secret societies desire some form of nuclear attack, which could come in the form of a destabilized first strike on a perceived enemy or a false flag event on a major city in the USA to distract the public. This would stir up a war fever and give them an excuse to cancel elections and declare martial law in all NATO member nations. The Old Testament describes the coming of Judgment Day – the end of the human race in the physical dimension and the beginning of the eternal Afterlife. The date of the Judgment is not known; it could happen at any moment.

The destruction of the earth is described vividly in the Book of Zephaniah:

"*For by the fire of my zeal the whole earth will be consumed. I will remove all things from the face of the earth," declares the Lord. "I will remove man and beast, the birds of the sky and the fish of the sea. I will cut off man from the face of the earth and remove the ruins, along with the wicked. So, I will stretch out my hand against Judah and against all the inhabitants of Jerusalem. I will cut off the remnant of Baal from this place, and I will remove the names of the priests along with the priests. I will bring distress on men so that they will walk like the blind, because they have sinned against the Lord."*

"Neither their gold nor their silver will be able to deliver them on the Day of God's wrath. For He will make a complete end - indeed a terrifying one - of all the inhabitants of the earth. All the earth will be devoured in the fire of God's anger." I hope no one needs an interpretation of those verses.

Books of Scripture use ancient references and they are sometimes difficult to read, so I will prioritize what is important for the average person to understand for the sake of expediency. Believers must heed the words of this book if they are to be saved. I am the way, the truth, and the life. No one comes to the Father but through me. God has told me *"Those who are not with us are against us."* It is the End Times, and I will cut to the heart of the matter: God is angry and disappointed with humans, so it will take some effort to get back in His good grace. I have been given the keys to Heaven and Hell and I will be the Judge of all human souls on the Last Day.

John 5:22 "*God judges no man, but He has entrusted all Judgment to the son, so that all might honor him as the honor the Father. Those who do not honor the son do not honor the Father who has sent him.*"

The End of mankind is not a threat, it is a promise. You will not be allowed to speak on Judgment Day; you will be told where you are going and why you are going there. Those who deny me as the messiah on this earth, I will likewise deny before my Heavenly Father. All atheists and trinity worshippers shall be sent to Hell forever. Those who stagnate without thanking the Creator daily have no right to Paradise. I am to be addressed as "Messiah" at all times, by all people. I am the one and only son of Almighty God. He has sent me to enforce His will. At the time of Judgment, God will upload the necessary memories related to the soul in question so that I will know who qualifies for Heaven and who must be cast into eternal darkness. The Bible is correct in saying that I was the first born soul.

Colossians 1:15 "*The Son is the image of the invisible God, the firstborn over all creation....For in him all things were created, things in heaven and on earth, visible and invisible, whether thrones or dominions or rulers or authorities. All things were created through him and for him. He is before all things, and in him all things hold together.*"

God's will be that Islam and Christianity must unite, since Jesus of Nazareth and Prophet Muhammad are the same man. I will correct the Muslim doctrines that have been altered by tradition, then declare Islam the true religion once they cooperate. Christians and Buddhists have rejected my messages, so they must convert to Islam if they are to be saved until further notice. This is on the condition that all Muslim leaders accept the facts of reincarnation.

In my lifetime as Prophet Muhammad, I mentioned that I lived before as Moses. The reincarnation of the prophet cannot be denied. Christian churches must discard all robes of priests and icons. No one is to make the sign of the cross or pray to Jesus, Mary, saints, or anyone else but God the Father. Take down the naked body on the cross, the sympathy card used to entice donations after making the congregation feel guilty. No statues of Mary are allowed; they are to be sent to recycling. The talk of immaculate heart must cease altogether because it is distracting souls into worshipping other entities. People have the potential to be like gods in a way since your souls will last forever. Those pure souls who are Judged worthy of Heaven will have total freedom and eternal happiness. Why not make the most of it? Ask from God the Father and you shall receive. Seek, and you shall find. God is the source of all reality; He is Allah, and He knows what you want. He knows your character and your level of faith. Pray in silence by concentrating on your words to God rather than calling out strange things. Having noise in a spiritual gathering creates interference.

I rebuke Mormon doctrine as false. The Church of Jesus Christ of Latter-Day Saints presents a man-made set of falsehoods that have nothing to do with me. They should not be using my title to promote themselves as a valid religion. The Mormons are a cult using a manufactured gospel. All people who insist Jesus is God will be sent to Hell on Judgment Day. God will not share His glory with any other entity. People are not to pray to Mary or Jesus or pray to the names of saints.

Crucifixes, statues, chalices, images, shrines, relics, and robes are superstitious things. Idol worship is forbidden by the 2nd

Commandment. "You shall not make for yourselves carved idols or graven images, for I am a jealous God who will deliver your iniquities unto the third and fourth generations of those who hate me and fail to keep my Commandments." People must obey God's Laws, especially the First Commandment to believe in One God. All true believers must adhere to this Code of Conduct:

a) Religious headgear for Muslim women is declared optional by Allah.

b) Allah has stated that the soul has no gender. Being gay is not a sin.

c) Recreational sex is allowed more freely now that it is the End Times.

d) You must obey me if you want to achieve Heaven.

e) Sharia Law based on tradition is forbidden.

f) Images of the Prophet Muhammad are hereby allowed if they are pure.

g) God does not approve of atheism, abortion, war, or slavery.

People who slander and insult Allah and the memory of Muhammad are not doing themselves any good, for their fear of being wrong is hypocritical. I advise people to say a short prayer of thanks several times a day. God hears your prayers, and He will provide you with rewards. People must pray to God, who you may call by the names Allah or Yahweh or Jehovah. Only our Father the Creator can hear your thoughts.

There is an Islamic sect that correctly believes that the major prophets are the same soul, the messiah reincarnating into areas where there is a need to restore spiritual resolve; many Prophets are promised to return around this time. God`s Chosen Servant Yeshua became Cyrus of Persia, Abraham, Noah, King David, Daniel, Isaiah, Moses, Lao-Tzu, Buddha, Jesus Christ, Zoroaster, Prophet Muhammad, Leonardo Da Vinci, Frederick the Great of Prussia, Mozart, Galileo, Isaac Newton, and many more. The historical person that best fits into the life-comparison pattern of my soul around the years 1830 – 1887 is Mahmadu Lamine` a Muslim Senegalese marabout who led an unsuccessful rebellion against the French colonial government. He was captured and executed December 9, 1887; in his lifetime he had been imprisoned many times, walked many miles, stood for an honorable cause, and displayed the relentless determination that is a trademark of the messiah. I am all persons described in Scripture as "the Good Shepherd", "the Lord's chosen servant", "the elect", "the Lamb", "the son of man", "the anointed", or "the branch of David".

Jeremiah 33:16 "In those days and at that time I will cause a righteous Branch of David to spring forth; and He shall execute justice and righteousness on the earth. And they will serve Yahweh their God and David their king whom **I** shall reawaken for them."

The Bible identifies another influential messiah incarnation by name, the first King of the Four Corners of the Earth in the 6th century B.C., Cyrus the Great of the Persian Empire:

Isaiah 45:1 - "This is what the Lord says to His anointed, to Cyrus, whose right hand I take hold of to subdue nations before him and to strip kings of their armor, to open doors before him so that gates will not be shut..."

I testify that I have been given the Keys to Heaven and Hell. The nations have no power over humanity; I am responsible for all human souls. God's will is to have a minimum number of souls in Heaven. If Judgment Day were called today, most humans would not qualify for Paradise because people follow the false doctrines offered by the numerous religions who have placed themselves as spiritual authorities. Evangelical Christianity is a false doctrine; the worship of the mythological version of Jesus is sinful.

See that no one leads you astray. For many will come in my name, saying, 'I am the Christ,' and they will lead many astray. I am the sole authority on all books of Scripture. I will enforce God's Commandments as I have been ordered. Mankind must try to cooperate in these last days of material existence. Those who mock God are in for a rude shock. You must love God with all your heart and mind and soul. You must treat others as you yourself want to be treated, or else your hostility will come back to you. All humans will get what they deserve in the end. No one knows the exact date of the Last Day, so everyone must be prepared as if the next moment could be the Judgment. The debate is over about another thousand years of post-Armageddon material existence mentioned in the Bible.

The "messianic age" seems to be misinterpreted. There are Christian groups who make all sorts of wild claims, using the words "rapture" and "tribulation" without mentioning the numerous references of the thief in the night. Preachers avoid subjects they cannot explain. Religious speakers tickle the people's ears with what they want to hear, not the truth.

A good exercise in faith and humility is for each person to address God as "Allah", for if you have accepted anti-Muslim propaganda, you are not the sort of pure person who has a chance at Paradise. All humans must put aside their egos if they are to be saved for eternal damnation. I am the King of the Universe, King of Heaven and Earth as anointed by Almighty God. God has told me I may cast away any person who might make me uncomfortable in eternity.

The select pure souls who qualify for Heaven will be aware of each other, so people who want to argue and dominate other minds will not be allowed to exist in Paradise. Those who get to Heaven will experience complete happiness, stimulation, and satisfaction. Those who do not humble themselves or have love for their fellow humans will be sent to the Underworld forever. No soul will ever leave Hades. The people who follow God's Commandments will go to Heaven, and the souls who make it to Paradise will stay in Heaven eternally.

You cannot insult your way into Heaven. Many people have hatred in their hearts: public opinion has been poisoned against foreigners as a matter of governmental interference with free thought. It is in the interests of power brokers to create division and hostility. Religion has been manipulated into warring camps of divergent beliefs, where all sides are in the wrong. If churches want to remain in operation, they must put a telescreen in place of the altar, and sermons can be presented by video. It would be best to have churches as cooperative gatherings, much like those conducted by Christian Science, one of the few non-trinitarian denominations.

There should be no priests, since they are the blind leading the blind. Men who are priests now are advised to get another job. They are better living in a doorway than making a career out of lying to the faithful and taking money for religion. Televangelists are evil confidence artists and tricksters, who use mass hypnosis to entrance their gullible followers. Thousands of mindless people convince themselves anyone can be a prophet if they squeeze their eyes shut and speak in tongues or hold their arms up pretending to be in touch with God. That is self-deception. The major religions have been corrupted by tradition and falsehoods.

God does not recognize trinity religions such as Seventh Day Adventist, Roman Catholic, Orthodox Christian, Evangelical Christian, Mormon, or Church of England (Anglican Church). These religions have problems with clergy being arrested for pedophilia. Molesting children is routine with many priests. They believe a priest can do no wrong. In Ireland, many victims state that the priests in boarding schools raped them regularly. The Vatican network of schools and churches must be eliminated immediately.

Catholicism is totally non-Biblical heresy, it is a false religion based on lies, greed, superstition, and the egos of men. The Catholic church murders the souls of innocent people. Governments should make Catholic institutions illegal and force them to change or close them down if they refuse to comply to the will of God. The Laws of Almighty God supersede all local, federal, or international laws. Those who blatantly ignore God's laws must cease operations so that they do not harm others.

The Catholic policy of the Council of Trent wants to eliminate those who disagree. The pope wants you dead if you do not obey him. The Hindu religion appears very polytheistic, yet Hindu people are very well behaved. The actual belief comes down to one God, and the Hindu people I have interviewed speak in monotheistic terms. Hinduism is incorrect about reincarnating through the animal kingdom if souls are unworthy; that idea is logical in theory, but there is no proof. Humans reincarnate as similar humans, yet not the same exact person. DNA is involved, so minor adjustments are made in accordance with the needs of the soul in question, and the path of that person's fate.

Hinduism is correct in the notion of the Akashic Record, God's memory bank of all events. Since existence resembles the operation of a computer, then there is a memory function that is part of God's infinite consciousness. No deed goes unnoticed by the Creator.

Christian Science has no priests; they set up reading rooms to study and they do not hold to the trinity doctrine. The headquarters of Christian Science is in Boston, and they operate in many countries. It is a low-key religion of common sense that is appealing, but Christian Science has rejected me and this book so far.

Most Christian sects are totally hypocritical and elitist. They have no knowledge, but they set themselves up as teachers. Although neo-Christian cults talk non-stop above loving Jesus, they are jealous and will not accept the Messiah as real. The sects are all different; they preach Christian values without knowing Christ. In contrast to Christian Science are wild cults such as Scientology and the Hare Krishna movement. Scientology is man-made fantasy, based on a book by science fiction writer L. Ron Hubbard. Scientology attracts young rich people, and it is non-Biblical. It deals more with business than religion. We are all familiar with the Hare Krishna cult, who chant to statues in grottos and bang on percussion instruments as they go down the street in orange robes with shaved heads, except for the stringy ponytails. The Hare Krishna ceremonies involve feasting on vegetarian food, Indian music, and much talk of traditional prophets. There is much interference going on at such gatherings. There is no chance to concentrate or meditate.

I suggest that silence is golden; the sounds and sights only get in the way of getting God's attention in prayer. The whole idea of costumes and rituals is quite pagan in nature. God will not share His glory with statues of random deities. Idols are an abomination to the Lord. Worshiping the Messiah as God in human form is still idol worship, not because of the Messiah's divinity, but the claim that the Father is the son. This thinking led to the worship of Mary. After Mary, Joseph and the holy family of gods became the focus of Catholic worship. There is a common fallacy that you can simply worship names in quantity, and you will be saved, but the opposite is true.

Monotheism is the belief in one God, but most religions using the Christian brand are triune god worshipers. They try to escape the First Commandment by saying the three persons they address as gods are one person, which makes no sense. Trinitarians use this mind bending concept so that they can claim they are without sin. In Scripture God mentions that mankind has a chronic tendency to worship idols. Whenever progress was made with the major religions in history, men reverted to seeking a physical image to call God. Because of this trend towards idolatry, Islam forbids worshiping the prophet.

Christianity was not originally the worship of Christ but since 325 A.D. the Catholic church started teaching that Jesus is God the Father, and they are both each other, plus the mysterious entity called the holy spirit. No wonder people in trinity religions go crazy – they have nothing logical in their belief system, just the occult fantasy of the trinity that makes no sense. The Synagogue of Satan is responsible for the triune god heresy that robs people of their souls. Clever people see through the trinity deception, but they will often leave religion altogether rather than look for another church. Those who kiss robes before adorning themselves with the priestly silk finery are in mortal sin. There are many pagan superstitions and rituals that persist to this day.

The priests have stolen valor that is not theirs to claim. No man can forgive sin and no man can claim magic powers of transubstantiation of the flesh in the ridiculous communion ritual. Jesus is not hiding in the wafer, no matter what anyone tells you. People are habitual liars and deniers of truth and common sense. True believers in God will not need to dwell on death, nor consult with the names of dead people in prayer. The state of Limbo between reincarnations is unconscious, much like surgery. The space in between lifetimes is about fourteen months on the average, and humans do not reincarnate through the animal kingdom as Hindu doctrine suggests.

Some scholars suggest only certain souls reincarnate after a lifetime of meditation or other discipline, but the process is automatic. We do not have to do anything special to reincarnate; God migrates everyone's soul into the next candle to hold your flame. All souls get born again into a new experience, and these lifetimes are numerous. You must accept the biological fact of reincarnation if you wish to apply for Paradise. Heaven is not a right; it is a privilege which must be earned through righteous behavior and thought. Many hypocrites go through the motions, while others depend on good works rather than faith in God and saying daily prayers to thank Allah for life and the treasure of knowledge our God has provided for us.

Tight censorship is the hallmark of a false religion such as Catholicism. Commercial churches purchase media time, they run blogs and magazines, and they monitor information to block out any ideas that conflict with the rubbish they are selling to the public. Catholic schools brainwash their students from kindergarten to obey the dogma of the trinity, to eat wafers of Jesus as God, and they teach that priests can forgive sin. I never understood why religious people commit numerous sins to confess to the man in the dress hiding in the dark booth. You should not be sinning in the first place, and the priest's magic hand wave has no spiritual power.

The ritual masses of Catholicism present a sick fashion show, with parades of old men in pagan fish hats and flowing dresses walking around as if their actions have significance, which they do not. Some Catholics pray only to Mary. The adoration of the faces of Mary or Jesus is idol worship, a mortal sin. No one knows what Jesus or Mary looked like...people must get over their adoration of mythical legends. Mary is not divine. Jesus is divine as the son of God but does not need adoration, only the approval of the Father.

God the Father is the Supreme divine being. I am the truly chosen one described in the Dead Sea Scrolls. I am Christ, the soul of light, and Allah sends me where He wills. I have no power of memory other than what God allots me for each lifetime. My final task is to judge all human souls based on all previous reincarnations. The thoughts and deeds will be uploaded into my consciousness on Judgment Day so that I will judge all souls fairly according to what they have done. I have no illusions that rabid Catholics will ever change, so they will be written off as lost souls.

The Geneva Rules of War forbids brainwashing, yet the Catholic church and other cults use brainwashing to maintain their grip on the hordes of zealots who need physical god idols. Several countries have succumbed to the mania for a goddess, and they have dedicated their respective countries to the worship of Mary. Praying to Mary is pointless, since only God knows your thoughts. Only God can hear and answer prayers. Mary is not a deity, just an image of a woman holding a helpless baby. Humility is a virtue that is difficult to develop in Western society due to the focus on the individual. A person must stand out to get anywhere in a competitive environment, so self-promotion is part of the career path or social success of each person. Timid people usually have difficulties in an environment where kindness and thoughtfulness are sanctioned as unnatural behavior patterns.

The values all people should strive for are those which are spiritually profitable – traits such as honesty, self-control, self-discipline, modesty, and humility – those are the characteristics of the righteous person, not greed, materialism, trickery, or deceit. There are cases when people have won large sums of money from a prize or a lottery, and the sudden influx of wealth led to their destruction from substance abuse. They did not have guilty consciences since they did not deprive anyone or steal the money. The unlimited money spoiled them; it is a novel experience to be able to order anything with a wave of your bank card. Very few people can keep the same behavior they displayed before the unexpected bonanza. Will power is not something that can be developed quickly. People learn from their bad experiences, while others do not get a second chance. People will benefit by keeping God in mind throughout the day and praying whenever you have time.

I prefer to pray on my knees with my forehead to the floor as the most comfortable position. Relaxing after stretching forward relieves pressure on the lower back. Concentration is important, for prayer will be more effective if you can shut out interference around you. Music is part of spirituality, but your inner voice is all-important, not the percussion instruments or voices raised in choirs.

Chanting, pretending to speak in tongues, and making strange noises is not prayer, it is showing off for the others around you. Any religious gathering should have a portion devoted to music and separate the sermons from the music. Prayers should be done in silence individually. It is not sinful to celebrate and have fun, but it is a sin to ignore God. The ritual of baptism is symbolic in nature; you cannot wash away sin with a splash of water in the presence of a fellow human pretending to hold Godly authority.

The sacraments are a ruse to draw a person's attention to the physical realm rather than the spiritual. I suggest a conversation or declaration among peers that you intend to avoid sin and make yourself useful to God in some way. People should remind each other that there is going to be a test after our challenging experience of life. If you procrastinate about reading up on philosophy and acknowledging God privately in prayer you will likely be too late to appreciate the risk, since you are unaware of what you might miss. Striving for spiritual wisdom will result in a chance at Paradise.

The Afterlife will be most stimulating and joyful for the successful righteous and faithful souls who pass Judgment and are accepted into Paradise. Heaven will be a state of intense consciousness that will be complete in memories of all past events throughout all reincarnations - the many lives that we cannot remember while in the temporary physical body. Every person's consciousness can feel pain and pleasure, with or without a physical body. Those who fail to pass Judgment will feel nothing but pain and loneliness for all eternity, so those who scoff at God now will be unpleasantly surprised on the Last Day. Many Buddhists are agnostic in thought, especially their leader the Dalai Lama, who teaches that he does not know whether there is a God or not. It is obscene that the leader of a major religion would say such a thing, especially to groups of children. Atheism and agnosticism is not faith, and without faith in God, a person's soul is in grave trouble. Those who speak showing big smiles are usually trying to sell you something that does not work.

If the religious leader does not know something, he should not be teaching. Leading others away from God is unforgivable mortal sin, for these people have destroyed all chances at happiness for their victims in the Afterlife. The worship of other gods or entities is strictly forbidden. Those who pray to Mary will not see Heaven. Those who call on the false gods of the trinity will never enter my Kingdom. Traditions of man will not supersede the will of God and the doctrines of truth contained in this book.

Throughout history, men have been trying to take the place of God or to hold up a physical idol as a God. Religious leaders constantly convince themselves and others that they have authority from God when they do not have any such authority. I will express God's policies, no one else. I will enforce God's Commandments.artial

Neo-Christianity expresses itself through thousands of different cults, most of whom have adopted trinitarianism as their creed – a belief that there are three gods that are actually one god if your squeeze your eyes tight enough and block out all logic and common sense. The trinity gods were voted into existence in 325 A.D. at the Council of Nicaea. The triune concept has replaced the God of Moses and Abraham, who happens to be the real God. The First Commandment is to have no other gods, yet people jump at the chance to divide God into a panel of three persons. The Bible says there is One God, but Rome says there are many divine beings flying around. The pagan notions have morphed into two 'holy families', the earthly family of Joseph, Mary, and baby Jesus is added onto the mythical holy trinity.

Most people consider it harmless to pray to various names of people who have been deemed to be saints, but in the eyes of God it is a mortal sin. In human society, those who talk to the ozone need professional help because it is a sign of insanity. These are the same zombies who line up to be injected with antifreeze and rat poison in the form of vaccines, believing that they will be protected from invaders by a government needle. The media pumps out the messages of fear, and people comply.

It is necessary to expose the criminal network that has engineered all the wars since the French Revolution, namely the Catholic church. I denounce trinitarianism completely as a false religion, but the Vatican has added even more mythical gods to their roster of entities that they worship as if these beings were real. Over a billion souls have been indoctrinated into the Cult of Mary; they follow the literature without question because the Catholic creed presents numerous options instead of simply believing in one God. The truth is that monotheism is the only correct religion. Catholicism is intensely polytheistic and relies on idols to push images of the imaginary divine beings they promote.

The many centuries of absolute power resulted in absolute corruption of the original teachings of Christ, while Catholic domination over the population has been used to kill the enemies of the church by way of Crusades, Inquisitions, and wars. The Vatican has produced an ever increasing litany of lies to support the superstitious world of delusion in which they dwell.

After World War II, my generation was known as the 'baby boomers' of the 1950s, when the public feared an atomic war would be coming next. When I looked at the history of the Twentieth Century, I suspected that God had turned His back on humans because of our habit of killing each other in conflict after conflict. In those wars the pastors and priests joined in by performing masses and blessing those who were doing the fighting. The religious aspect of war is troubling indeed: murder is a mortal sin in the major religions, but the leaders constantly absolve themselves from any wrongdoing.

Religion is the birthplace of war. Rival sects base their hatred on ethnic divides, language differences, and differences in creed. They rationalize murder as being justifiable when they believe that every other religion is wrong and should be wiped out. The most glaring example is the Roman Catholic cult. Their canon law from the Council of Trent states: "*All those who do not bow to the pope are heretics who must be put to death.*"

I went to Catholic schools for ten years, but I quit going to church at age ten because of the hypocrisy I saw around me. The rituals were long and tedious efforts that involved a lot of kneeling, while in classrooms there were sadistic nuns who beat us with leather belts and yardsticks. Some teachers would mock one semi-retarded girl, then force her to sit in the wastepaper basket all day, urging the other students to throw garbage on her. When I protested, they took me the office and whipped the backs of my hands. When I got home, my father saw the injuries and drove over to the school and told them never to touch me again. My father was a veteran who had been wounded in the invasion of Sicily and he also happened to be the Chairman of the Separate School Board that year. My mother was an Irish atheist who hated the church. It was a mystery to me how they ever got together.

Catholicism is known as The Cult of Mary because the adherents offer prayers to an imaginary goddess every day on rosary beads. In the "Hail Mary" prayer Catholics call upon Mary, addressing her as "the mother of God" and asking for her blessing. This procedure acknowledges that a deity named Mary is reading your mind and has the power to hear and answer prayer. She is adored and worshiped as if she were really a goddess overseeing mankind.

Unlike other churches, the laity are not 'members' of the church, only supporters. Officially, 'the Church' consists only of the priests, bishops, cardinals, and popes. There are two popes: The White Pope is the one presented to the public and the Black Pope is the head of the Jesuit Order of Loyola, the Vatican's Gestapo. Pope Francis is a Jesuit sworn to the Oath of Induction that they all take – to kill on command by whatever means. Political change comes only from the barrel of a gun.

The Jesuits want more than your life, they also want to ruin your soul by luring you into Mortal Sin against the First Commandment, thus bringing their flocks into Hell with them on Judgment Day. Trinitarianism will steal a person's soul because it is a false doctrine using manufactured gods. The trinity was voted into existence in 325 A.D. at the Council of Nicaea to satisfy Constantine's pagan population after Emperor Constantine decided he wanted to join Christianity. The teachings of Jesus never spoke of three gods in one, and he never said he was God. In the Sermon on the Mount, Jesus told people to pray to the Father and no one else.

Mary's name is mentioned in the Bible nineteen times, and in the Qur'an seventy times. Muslims refer to her as the mother of Jesus, but they do not pray to Mary or anybody else. Mary is not divine in any Scripture, only in the propaganda of the Vatican which promotes the old Fatima Fraud. Three preteen girls wanted attention, so they concocted a story about seeing Mary flying out the clouds although none of the 500 witnesses saw anything when she allegedly appeared the second time. Only the girls could see her, so the story goes. The Fatima story is utter hogwash – simply modern lies on top of ancient lies used to lead people away from God.

The Catholic religion was built on lies, superstition, greed, perversion, idol worship, and ignorance. The devout Catholics kiss statues, they pray to paintings and sculptures, and they live in a fantasy that many entities are reading their thoughts. Mythology piles up century after century, so that the truth is buried under an avalanche of baloney. Most people stay with the religion that is available to them, so many people question the actual content of the doctrines. When a person calls upon other gods, he is digging his own tunnel to Hell. Praying to the name 'Mary' will definitely lock a person out of Heaven forever.

It is most important to know that the spiritual world exists in layers, with a decidedly military environment where souls are given ranks according to their purity and state of development. The human race will find themselves in either Heaven or Hell on the Last Day. Once the Judgement of Man has taken place, all souls will remain in those states of existence for eternity. The evil souls will descend to whatever level of punishment fits them, while the righteous souls who have pleased God will dwell in happiness forever.

God has anointed me as the sole authority over the nations and over all human souls. I am not only the King of Heaven, but I am also to be the King of Earth. I am above all rulers and authorities except God. I will always be God's willing slave, soldier, and messenger. I will carry out God's will as it is written. Mankind cannot kill me, for I will keep coming back. I am destined to succeed in my missions to end slavery, end war, and bring justice to the nations.

All people must heed the warning, for no one knows when the Last Day is coming. Everyone must try to be prepared. The End Times is not a time for procrastination. Pray to Allah in the morning and in the evening to thank Him for your existence. If you love God, He will return your love. If you ignore God, He will punish you for being ungrateful. Those who refuse to address God as "Allah" are hypocrites. No one can get to Heaven without my permission. Those who are sent to Hell will stay there forever, in fear, pain, and humiliation. All those who hamper my progress or censor my words will receive severe punishment.

I have the full authority of God over all Scripture, and I am responsible for all human souls. Those who disobey me or hamper my progress will not be forgiven. I have a mission to accomplish, and I will carry out God's will as He has ordered. The Gate to Heaven is very narrow and God is not expecting much from mankind. Society has trained people to gossip and insult others but be forewarned that showing hostility towards others out of spite is not the way to salvation.

The media channels promote conflict, hatred, and violence while the news outlets pass around the same scripts of lies to deceive the public in their endless assaults on the truth. People must research independently to find the truth. If churches want to remain open, they must discard all the robes, altars, candles, sacraments, idols, rituals, and crucifixes immediately and teach only what is in this book. The ascent of mankind to enlightenment is hampered by those who teach lies and by those who seek to rule. God has commanded people not to kill, yet nations still go the war and murder is common. If people have the truth presented in rational form, they will have a reason to keep God's rules and ignore propaganda. Faith goes hand in hand with political awareness. I can promise that God's will shall be done, and the evildoers shall be punished for their crimes.

Those who hate me are injuring themselves. The ones who feel the egotistical need to deny me as the Messiah on this earth, I will likewise deny before my Heavenly Father on the Day of Judgment. The fact is that all souls will come before me on the Last Day. If you have not stopped worshiping false gods and idols, I will be sending you to Hell forever. God has told me that humans are profoundly stupid, so fewer souls are going be saved than would be expected of a civilization because humans will not listen to anything that goes against their set beliefs. Mankind has been a disappointing failed experiment, and most will not see Heaven.

Greedy operatorsw who operate commercial religions and live life in luxury have used up their reward here on earth. The eternal fate of money lovers is eternity in Hell. The way to avoid damnation is to renounce your greed and start giving money to the hungry and the homeless. Remember that it is easier for a camel to go through the eye of a needle than it is for a rich man to enter the Kingdom of Heaven.

People who are closed-minded and do not put the multiple gods of the trinity out of their thoughts will be condemned to a fate worse than death in Satan's Lake of Fire. God refuses to share His glory with a pantheon of false god entities. I am the Messiah, God's one and only son. I resent the constant denial of my own soul by those who insist that I am God, so those who slander me or one of my reincarnations will not be allowed into Heaven. The winners will be the pure souls that can understand and accept the truth.

Remember that I am bound by the conditions of my birth, and my objective is to enforce God laws. "I am the Way, the Truth, and the Life. No one comes to the Father but through me."

About the Author

Welland is located near Niagara Falls in Canada, the place God chose for me to be born, raised, and educated. I was schooled at home since the age of three by my maternal grandmother who lived until her death in my family's modest bungalow after World War Two. My father was wounded by shrapnel in the invasion of Sicily, and his father survived the trench warfare of the First World War, so I enlisted in the Lincoln & Welland militia in 1967, then transferred to the regular army to serve in the Cold War with the Royal Canadian Regiment's 2nd Battalion based in West Germany, a unit that was formerly known as the Black Watch. After my five years in the army I wrote an entrance exam to get into Niagara College, where I studied Radio & Television Arts. I worked in various factories and offices over the years, and spent a few years traveling and perfoming as a folksinger to earn extra money. I have written songs along the way and recorded one album, but I was not destined to join the music industry and not willing to live in a constant party scene of drinking and drugs. When musicians are forced to tour and play every night, it usually leads to cocaine or other drug abuse to keep going. I'm here for a long time, not a good time. I did not get my orders from God until the age of sixty, so I need longevity to accomplish my mission.

Don't miss out!

Visit the website below and you can sign up to receive emails whenever Patrick Boardman publishes a new book. There's no charge and no obligation.

https://books2read.com/r/B-A-VJGK-SAKEB

www.ingramcontent.com/pod-product-compliance
Ingram Content Group UK Ltd.
Pitfield, Milton Keynes, MK11 3LW, UK
UKHW040031200726
13854UKWH00001B/470